Airline Governance

Anyone becoming a company director faces a steep learning curve; this book will give every director and especially one joining the board of an airline, a head-start on the process.

Airline Governance: The Right Direction will help existing directors, those who have been newly appointed and those 'in waiting' in a company's management. This book reviews the fundamentals of corporate governance and puts them into the context of guiding, directing and managing an airline, and also complements the discussion of accounting and finance in its sister book *Airline Management Finance: The Essentials*. The detailed review will give directors confidence to make decisions on governance matters, avoiding a 'tick the box' approach and focusing on what is important. This book not only gives directors a comprehensive introduction to good governance, but also discusses the application of the principles of governance for an airline at various stages of its development so any changes can be made at the right time.

Understanding corporate governance not only helps directors, but also an airline's senior and junior management, because the considerations around matters such as 'conflict of interest' apply to all decision-makers in the organisation. Understanding and applying good governance does not guarantee success, but it surely helps in achieving it.

Victor Hughes is a Chartered Management Accountant and a Fellow of The Hong Kong Institute of Directors who spent most of his career in asset-intensive industries and was Finance Director on the board of Cathay Pacific Airways Ltd.

Managing Aviation Operations
Series Editor: Peter J Bruce
Associate Editor: John M C King

The purpose of this series is to provide a comprehensive set of materials dealing with the key components of airline operations. To date, this innovative approach has not been evident among aviation topics and certainly not applied to operational areas of airlines. While more recent works have begun, in brief, to consider the various characteristics of operational areas, the Managing Airline Operations for Aviation Professionals series will expand coverage with far greater breadth and depth of content.

Airlines are devoid of specific topic knowledge in ready-made, easy-to-read, creditable resources. Tapping into industry expertise to drive a range of key niche products will resource the industry in a way not yet seen in this domain. Therefore, the objective is to deliver a collection of specialized, internationally sourced and expertly written books to serve as readily accessible guides and references primarily for professionals within the industry. The focus of the series editors will be to ensure product quality, user readability and appeal, and transparent consistency across the range.

Airline Management Finance
The Essentials
Victor Hughes

Airline Operations Control
Peter J Bruce and Chris Mulholland

Airline Governance
The Right Direction
Victor Hughes

For more information about this series, please visit: www.routledge.com/Aviation-Fundamentals/book-series/MAO

Airline Governance
The Right Direction

Victor Hughes

LONDON AND NEW YORK

First published 2021
by Routledge
2 Park Square, Milton Park, Abingdon, Oxon OX14 4RN

and by Routledge
52 Vanderbilt Avenue, New York, NY 10017

Routledge is an imprint of the Taylor & Francis Group, an informa business

© 2021 Victor Hughes

The right of Victor Hughes to be identified as author of this work has been asserted by him in accordance with sections 77 and 78 of the Copyright, Designs and Patents Act 1988.

All rights reserved. No part of this book may be reprinted or reproduced or utilised in any form or by any electronic, mechanical, or other means, now known or hereafter invented, including photocopying and recording, or in any information storage or retrieval system, without permission in writing from the publishers.

Trademark notice: Product or corporate names may be trademarks or registered trademarks, and are used only for identification and explanation without intent to infringe.

British Library Cataloguing-in-Publication Data
A catalogue record for this book is available from the British Library

Library of Congress Cataloging-in-Publication Data
Names: Hughes, Victor, 1941– author.
Title: Airline governance : the right direction / Victor Hughes.
Description: Abingdon, Oxon ; New York, NY : Routledge, 2021. | Series: Managing aviation operations | Includes bibliographical references and index.
Identifiers: LCCN 2020022119 (print) | LCCN 2020022120 (ebook) | ISBN 9781138610712 (hardback) | ISBN 9781138610729 (paperback) | ISBN 9780429465635 (ebook)
Subjects: LCSH: Airlines–Management. | Corporate governance.
Classification: LCC HE9780 .H84 2021 (print) | LCC HE9780 (ebook) | DDC 387.7068/4–dc23
LC record available at https://lccn.loc.gov/2020022119
LC ebook record available at https://lccn.loc.gov/2020022120

ISBN: 978-1-138-61071-2 (hbk)
ISBN: 978-1-138-61072-9 (pbk)
ISBN: 978-0-429-46563-5 (ebk)

Typeset in Bembo
by Newgen Publishing UK

Contents

Acknowledgements vi

Introduction 1

1 Corporate governance and how it works 3

2 The Board of Directors and how it works 16

3 Organising the board 34

4 Strategy 50

5 Reporting 67

6 Changing corporate governance with growth 85

7 Subsidiaries and related companies 99

8 Reviews 112

9 What comes next? 127

Glossary 143
Index 159

Acknowledgements

During the development of producing this book, as well as its sister publication, *Airline Management Finance: The Essentials*, many people shared their expertise, experience, views and suggestions on corporate governance in general and corporate governance in airlines in particular. I enjoyed the discussions I had individually with, Willy Boulter, Derek Cridland, Ray Fung, James Ginns, Rupert Hogg, Alastair MacAuley, Martin Murray and Ron Taylor and am grateful for their observations and advice. Their input went into the grinder of my mind, but the output is entirely mine as are any errors or omissions.

Particular thanks to Peter Bruce and John King for their advice and help in developing the two books and for devoting so much of their time to reviewing and making suggestions on the draft texts.

Everyone I had contact with in the publisher, Routledge, were helpful, sympathetic and positive and helped smooth the process of producing both books.

My thanks to my wife, Helen, for her patience and support while I was writing the books, also for reading and commenting on each draft; tireless determination. Many thanks.

Introduction

Being appointed to the board of a company is often thought of as an honour and the ultimate career accolade crowning a successful career. In many ways this reaction is correct, it is an honour to be given the chance to guide a company towards success, but with the honour comes new responsibilities. Companies are an important part of a nation's economy; having successful companies will mean the total economy can be successful.

Usually directors arrive on a Board of Directors via one of two routes, promotion from within the company, often following a successful career in a specialist or technical department e.g., Marketing or Flight Operations, or by appointment from outside the company of a person with knowledge, experience or contacts which will be useful to the company. In both cases, after the initial feeling of pleasure from the invitation, a couple of questions will quickly spring to mind, 'what is involved in being a director?' and 'how does the board run this company?'. This book seeks to answer both questions and to give newly appointed directors the confidence to do their new job well.

It may seem strange that a newly promoted director or one with experience on the board of another company might ask 'how does the board run this company?', because an assumption could be made that most companies will follow the same system and approach. This is not completely true, the boards of most companies use the same principles to guide and administer their companies towards success, but they take those principles and apply them in the way that is appropriate and beneficial to the individual company's industry, circumstances and size. Hence the principles of governance don't change, but the way they are applied is and should be related to the needs and circumstances of the company; there is not one template that fits all companies.

When a director first joins a Board of Directors there is a lot to learn, even if the individual has served on other Boards of Directors. For an executive director, they need to understand that being a director means they have another, separate, job and their function is not to represent the area of the company they manage, but direct and guide the company as a whole. All Boards of Directors have a general duty to make their decisions in the best interests of the company as a whole, not one particular area or department. A non-executive director has to understand the industry and the objectives of the company and how it operates, so they can offer their experience and expertise to help the company meet those objectives. Either way there is a lot to learn; even a non-executive director moving from one airline's board to another airline's board will need to understand the differences in the airline's objective and operations.

Being a director is a professional job in its own right and each director needs to be able to allocate sufficient time to directorship matters so they are able to do the job well. There is also the subject of education, initial and continuing learning, about the company and the corporate governance. Directors need to understand what is involved in being a director and should keep their knowledge base up to date as laws and best practice recommendations change over time.

There seems to be an assumption that 'good corporate governance' is a project or approach which is the monopoly of big well-established companies and this is simply not true. In many ways, smaller companies and start-ups can derive more benefit than larger companies by their directors understanding and applying the principles of good corporate governance. Introducing workable governance during the early stages of a company's development will give the company credibility with investors and lenders and sow the seeds of a system which will serve the company well as the company grows. There is no significant difference in good governance principles between a small company and a large company, the only differences are scale and complexity. A good system of governance will serve all sizes of company well.

Different companies and industries have different priorities at different times. For example, in an airline dealing with the subject of 'safety' requires a different approach to 'safety' in (say) a trading company. Both may give safety the same priority, but the definition of safety and how the management deal with the issues will be different. The principles are constant but their application differs to reflect the different circumstances of the different industries and companies.

There is not one universal ideal system of corporate governance. Directors need to understand the principles of good governance, how the company is currently operating and find the best, most workable, way for the company to apply those principles to support the development of the company as circumstances change. The board must always be aware that as the company grows and changes, the company's good governance practices will also need to change to continue to help the company.

Good corporate governance reaches into the whole organisation, it is not just a matter for the directors. The policies and practices which the Board of Directors approve will apply to everyone within the organisation and some may affect those outside the organisation. An everyday example is, the ways to deal with potential conflicts of interest apply as much to a manager issuing a tender for supplies as they do to a director agreeing to the terms of a loan.

Directors who understand the principles of corporate governance are able to see the benefits to their company, of whatever size and stage of development, and decide what is the best way for them to implement the principles and realise the benefits. Having good governance practices does not guarantee success for a company, but it will certainly help.

1 Corporate governance and how it works

What 'corporate governance' means

'Corporate governance' is a term which is frequently used in news reports and conversations, so it may seem strange to start this book with a discussion of what the term means. There are a lot of definitions of corporate governance, all of which seek to describe it clearly. There are many eminent authors of definitions for 'corporate governance', including The World Bank, Organization for Economic Cooperation and Development, various Institutes of Directors, academics and government committees (e.g., the United Kingdom's Cadbury Committee).

Producing one definition of corporate governance is difficult for two main reasons. Firstly, governing a company means people managing people, so it is a human activity and human beings are individuals and act and react in different ways. The second factor is that shareholders' expectations of how a company should be managed are changing over time, frequently in response to some major commercial problem. In addition, society in general also has expectations of how companies should be managed and how interest groups are able to influence the thinking of governments on corporate governance. It follows that as the circumstances surrounding corporate governance are not fixed and there cannot be an absolutely complete, and etched in stone, final list of a director's responsibilities. The best that can be produced is a list of what is currently accepted as being the responsibilities of directors. In such a fluid environment producing a definition (i.e., a clear complete statement of the meaning) for corporate governance, is difficult. For this book, the term 'corporate governance' will mean 'the rules, laws, policies and practices which govern the operations of a company'.

Directors and shareholders

In a company the ownership and the operation of the company are separate. This separation is fundamental to the way in which most economies work in the world and arose from the landmark case of *Salomon v. A Salomon & Co Ltd.* (which was decided by the United Kingdom's House of Lords on 16 November

1897 with the ruling that the creditors of an insolvent company could not sue the company's shareholders to pay up outstanding debts owed). This case concluded that a limited company was a separate entity from its shareholders. Essentially the shareholders own the company, although even this statement can be debated because some lenders may have a prior charge over all or some of the company's assets, and the shareholders appoint directors to run the company for them. This separation produces the need for an approach which allows shareholders to be assured that the company is being well run, however that is defined, and is meeting its objectives, all without the shareholders interfering with the day-to-day operations of the company. The solution to this need is the requirement that the company's directors periodically and regularly report to the shareholders on the result of the company's operations and the current financial and business position of the company. In addition, the directors are required to run the company for the benefit of the shareholders. In general, this means they must make decisions in the best interests of the company and exercise independent judgement. They must also use reasonable care, skill and diligence in all matters relating to the company. When someone is managing other people's money, which is essentially what directors are doing, it is not unreasonable to expect them to exercise common sense and to periodically report on how the money has been used, what has been achieved with the money and what the future plans are. This duty to act in the best interests of someone else is often referred to as a 'fiduciary duty'.

Fulfilling this fiduciary duty does not involve taking instructions from the shareholders or constantly wondering what the shareholders are thinking about the company, but managing the affairs of the company for the long-term benefit of the shareholders. Later there will be a discussion about the relationship between the Board of Directors of a company and the company's stakeholders.

Directors and operations

The first question to answer is why is a Board of Directors needed at all? Adopting the approach of having a Board of Directors, which is really a committee charged with running the company on behalf of its shareholders, assumes that a committee of individuals, each of whom give independent advice based on their knowledge and experience, will make better decisions than one individual or a group who are involved in the ownership of the company.

Up to this point there has been mention of two groups involved in a company; the company's shareholders and its Board of Directors. But the directors do not operate the company; they guide it strategically and monitor its financial and operating performance but others will actually operate the company day-by-day and do what is necessary to achieve the objectives set by the board. Clearly then there is a relationship between a company's directors and the operating staff.

The directors are responsible for defining the company's strategy and monitoring its operations, whilst the operating staff with their more detailed knowledge of the company's operation are charged with implementing and achieving the strategy while following the policies agreed by the board. These two groups, the directors and the operators, need to work closely together and have a full understanding of each other's responsibilities and problems if the company is to be a success. If the operating staff do not understand what the company's strategy is and what the agreed tactics to achieve the strategy are, it is unlikely that the strategy will be achieved. Not understanding the strategy and tactics could mean that operating decisions actually make achieving the company's goals more difficult. Similarly, the directors need to understand what operating resources are needed to achieve the strategy and any operating constraints that need to be overcome. There needs to be good communication between the board and the company's operating management. To some extent the need to have good communication between the board and the company's management is helped by including on the board one or more 'executive directors' (i.e., directors who have executive or management responsibilities within the company), but at best this can only be part of the communication process and is not a substitute for an established method of communication between the board and the company's management.

There is a system of dual Boards of Directors which is used extensively in Europe and acknowledges the difference between direction and management. Under this organisation a company has two boards, a Supervisory Board and a Management Board. This arrangement is generally called a 'bicameral' board or system.

The Supervisory Board is responsible for:

- Agreeing strategy.
- Interrogating and approving plans produced by the Management Board.
- Deciding on company policies and good practices.
- Monitoring the performance of the company's operation against approved plans and strategy.
- Maintaining relations and communication with shareholders and other parties.

The Management Board is charged with:

- Running the company day-to-day using policies and good practices approved by the Supervisory Board.
- Implementing the strategy agreed by the Supervisory Board.
- Preparing short- and long-term plans to be reviewed by the Supervisory Board.
- Reporting on progress to the Supervisory Board.

Other responsibilities may be added to the mandate of either board.

The bicameral board has its advantages and disadvantages, but as a generalisation most economies prefer companies to have just one Board of Directors. This is particularly so in most Common Law jurisdictions where companies have a 'unitary board' with all of the responsibilities resting with one group; the Board of Directors. Even with a unitary board it is not uncommon for a company to have a Management or Executive Committee comprised of executive directors and senior managers who are charged with similar tasks to the more formally titled Management Board. With this approach the responsibilities of the company's Executive Committee are defined and formally delegated by the board to the committee.

Perhaps the clearest statement of the allocation of duties between a Board of Directors and a company's Executive Committee can be found in the agreements between the shareholders of a joint-venture between two or more companies. Each party in the joint-venture wants to be clear who does what and to have clarity on where the limits of authority are for decisions on such matters as capital expenditure, executive remuneration, buying and selling investments and all the other key matters.

When these three layers are looked at together, they link the shareholders to the operation of the company. Whichever approach is taken, the Board of Directors (or Supervisory Board) remains responsible to the shareholders for the governance of the company.

The role and duties of a Board of Directors and an individual director

Having established that the directors are responsible for a company's corporate governance, it is appropriate to discuss the role of the Board of Directors and the duties of a director.

Given the discussions and thought applied to corporate governance over many years, it might be assumed there would be a generally accepted complete list of duties for the board, but this is not the case. The essential problem is that while the general scope of the role and duties of a Board of Directors can be reduced to writing, the definition of just what each duty means is liable to develop and change. Ten or twenty years ago some matters which now are high on a board's list of considerations were probably seen as less important such as, for example, the possible effects on a business of artificial intelligence, or the need to protect personal data, and there are many more.

A useful general description of what the Board of Directors has to achieve is 'to run the operations of the company for the best interests of the shareholders'. This is such a general statement that most directors, not unreasonably, require some further idea of what their responsibilities are. This could be to:

- Determine and agree a long-term strategy for the company and to define each element of the strategy so that progress towards achieving them can be measured and monitored.

- Establish operating tactics designed to achieve the agreed strategy, monitor their effectiveness and modify as necessary.
- Identify, secure and allocate the resources (e.g., staff, funding), needed to achieve the long-term goal.
- Agree the way the company should be organised, review the performance of the organisation and agree changes as necessary, including the appointment of the senior executives.
- Establish sets of values to guide the organisation in its day-to-day operations.
- Agree a clear definition of the duties of the company's management.
- Regularly monitor the company's financial and operating performance and the progress towards achieving the company's strategy, comparing the outcome with past results, approved plans, competitor's results and market conditions.
- Identify the risks facing the company which might frustrate the achievement of the company's strategy and ensure plans are developed to handle the risks.
- Regularly report to shareholders and stakeholders on the state of the company and its progress on reaching its strategy.
- Ensure there are procedures in place which confirm that the company complies with all legal requirements.
- Monitor the company's cash position to ensure the company remains 'solvent'. That is, able to pay its debts when they fall due for payment, not all debts at one time – just those which are due. This is a very significant duty and one which is well documented. Failure to ensure a company remains solvent significantly changes a director's duties and focus. When a company is insolvent the directors have a duty to put the creditors' interests first and to seek to remedy the situation. If the directors permit a company to continue to operate knowing that it is insolvent, the directors may not only become personally liable for the debts incurred, but be prohibited from being directors now and in the future. If it appears that the company will become insolvent the directors should get professional advice very quickly.

Each of these elements can further be broken down into greater guidance, and into more detail, but even this list will not be fully complete and final, because the law does not have a fully comprehensive definition of a director's duties. Guiding and monitoring the operation of a company is essentially a human activity and this makes it difficult to define and regulate. Whenever there is a major business collapse or commercial disaster, there will be questions as to the extent to which there was some fault or problem in the business' corporate governance. The conclusion and lessons learnt from the problem generally produce a further clarification of a Board of Directors' role. The definition is a continuing process. Some people see this lack of definition as a major risk to themselves and their reputation hence decline to become directors. Although there are very few cases brought against directors for failure in their duties, some people still see the risk to their reputation as significant. Being a director does involve accepting significant responsibilities.

To fully understand what is expected of them when they join any Board of Directors, a director should get a copy of the latest guidance from their local Institute of Directors and study it, then read books on corporate governance in general. Fortified with this accumulation of knowledge and opinion a director will have a better understanding of what is required of them.

The duties of a director are easier to discuss and define, because they are frequently listed and defined in some form of Companies Act, which is the legislation governing the operations of all companies in a jurisdiction. Although there may be variations between jurisdictions the key duties are to:

- Act honestly and in accordance with the law.
- Exercise their powers for the benefit of the company.
- Promote the success of the company.
- Ensure that complete, up-to-date books of account are kept.
- Use due skill and care when making decisions for the company.
- Avoid conflicts of interest. This involves more than just not taking any personal advantage, but is also a prohibition against a director making a secret profit on transactions with the company.
- Consider the position of employees.

There is also a list of legal duties, which also varies between jurisdictions, and includes:

- The keeping of the records which the company is required to have (e.g., a register of shareholders, a register of directors and Company Secretaries) and have the records up-to-date and ready for inspection. Some jurisdictions even specify the format of the registers a company is required to keep.
- The reporting of the information required by the local government office responsible for dealing with companies, often called the 'Companies Registry'. Examples of the information which should be reported are, the details of the current directors, the appointment of new directors and their details, the removal of the company's auditors.

Frequently these duties are carried out by the 'Company Secretary' who is the company's officer responsible for keeping the company's legal records (e.g., records of directors' and shareholders' meetings, and a register of shareholders).

Reporting to shareholders and stakeholders

Shareholders

It would not be reasonable or sensible for a shareholder to give money to a company to use in a business or for them to buy shares in a business from someone else without knowing what the business has achieved in the past and what its future prospects are. In addition, when the shareholders put more money into the company, they will want to know what the new money will be used for. The law generally requires that each company regularly report to its shareholders on the state of the business, and this is frequently called the 'annual report' or 'statutory report'. The requirements on what should be included in the report to shareholders varies from country to country and is normally mentioned in the law; where the company's shares are traded on a stock exchange, the exchange may also lay down additional requirements. The common minimum requirements are for the report to include:

- Statement of the financial result for the period, often called a 'Profit and Loss Account' or 'Statement of Profit and Loss and Other Comprehensive Income'.
- Statement of the current financial position, frequently called a 'Balance Sheet' or 'Statement of Financial Position'.
- A summary of the company's cash flows, both inflows and outflows, called a 'Statement of Cash Flows'.
- A commentary on the business and the financial statements together with an indication on the company's future prospects.

The calculation, content and presentation of these financial statements are generally governed by local or international 'Financial Reporting Standards'. The major exception to this general rule is in the USA where a different set of Financial Reporting Standards is used. There are plans to harmonise these differing reporting standards in the future. The international standards are often referred to as 'IFRS' ('International Financial Reporting Standards') and are produced by the 'International Accounting Standards Board' (IASB). The international standards are frequently adopted by a country's national accounting body and issued locally as national Financial Reporting Standards. These standards are kept under review by the IASB and are frequently revised based on the experience of the companies using them.

The commentary by the directors on the current state of the company's business, the general business environment, the challenges the business faces, the opportunities that are currently seen to exist and the near-term and longer-term prospects are all very important, not only to the company's shareholders, but also lenders and suppliers.

The content of these discussions or reports has developed and expanded a great deal in recent years and now not only deals with the financial figures, but also the company's impact on the environment and how it is seeking to minimise any harm. There will also be comment on social matters like the contribution the company makes to society, locally and generally. This trend of more extensive reporting on non-financial matters is likely to continue. All directors need to keep up to date with developments in company reporting to ensure the company meets any new requirements and is able to report to a high standard.

The shareholders, although often thought of as one group, are not all interested in the same aspects of a company. Some shareholders may be holding shares because of the company's 'dividend' (i.e., the part of the company's profit paid to shareholders). Other shareholders may be interested in the value that their shares represent expecting that it will increase. Shareholders can be individuals or professional managers investing on behalf of others. Whether individuals or investment managers, some will be interested in the short-term results and value of the company, but others in the company's long-term prospects. The reasonable needs of each of these types of shareholders need to be considered by directors when they decide what to include in the company's annual report.

The company's directors meet the shareholders periodically, a minimum of once a year. It is usually compulsory for there to be one meeting a year and it is often called the 'Annual General Meeting' or 'AGM'. Shareholders are not obliged to attend the AGM, but all directors must. The AGM is a good forum for directors to explain the company's financial and operating results and business prospects. Shareholders have the chance to question any aspect of the business. The only practical restriction is that directors are not required to reveal any commercially sensitive information.

Stakeholders

For very many years businesses only concentrated on reporting the company's financial position and then only to its registered shareholders. Over time it became clear that there are others who are interested in a company's business:

- Potential investors who want to understand the company and how it operates.
- Financial institutions, both those who currently deal with the company and those considering starting a financial relationship.
- Suppliers of goods and services, who want to understand the company's financial standing and plans for the future.
- Company staff and staff associations.
- Government organisations.
- Organisations interested in the company's effect on some aspect of the economy or society.
- For an airline, the safety regulator.

The list is impressive and there are more that could be added. The general term for those who are interested in the company's activities is 'stakeholders'. There is a demand for information and the company's annual report can be developed into an important communication document with all stakeholders.

Except where stakeholders are also shareholders, there is not, at present, a requirement for a company's directors to meet groups of stakeholders to discuss the company's current and future position. It is, however, not unusual for the Board of Directors or an individual director to periodically meet financial institutions to review the company's position. Similarly, there may be meetings with investment managers.

A wise Board of Directors will also establish an effective method to communicate the information which is in the company's annual report to the company's staff, as they are an important audience. Staff need to know what the company has achieved and what the priorities for the near-term and long-term are. There are a range of options for discussions with staff, from face-to-face meetings with directors to written and electronic communications, and probably a combination of all of these.

Whether there are face-to-face meetings or written communications with groups of other stakeholders depends on the approach directors decide to take and whether there are any key issues which need to be explained.

Reporting in general

The information and discussion in the annual report should be comprehensive, easy to read and understand, and also expressed in a straightforward way. The objective is to communicate the key information to the reader, not overwhelm them with large quantities of data. Care needs to be taken when deciding what information about the company's future and plans to include in reports to shareholders and stakeholders. The information should give the reader an indication of what the future may hold for the company, but there is no obligation to reveal any information which might jeopardise that future. There is no need to mention any business secrets, commercially sensitive information or plans for acquisitions, if mentioning the information might prejudice the company's position.

The subject of financial reporting is discussed in more detail in the sister book *Airline Management Finance: The Essentials*.

Help for all sizes of airline

There may be the impression that the discussion so far, which has touched on strategy, bicameral or unitary boards, stakeholders, etc., is only of interest to large publicly quoted companies with hundreds of staff and is not relevant or important to a small airline or one just starting up. This is not the correct impression as adopting the principles of good governance can be a great help to smaller companies, but how those principles are applied to a smaller operation may be different. Also, the way the principles are implemented will probably need to change as the airline develops. There will almost certainly be a need to make changes if the Board of Directors decides to seek a quotation on a stock exchange. Stock exchanges tend to have their own additional reporting and governance requirements, but that comes later in this book.

Governance for smaller companies raises again the important point in this book, that directors need to understand the principles and objectives of good governance and apply them in a way that is relevant to the company in its current circumstances. A couple of examples may help to explain this point:

- Earlier in this chapter one of the duties of a Board of Directors was identified as to 'determine and agree a long-term strategy'. This may not seem to be an easy thing to do for (say) a group of experienced airline managers seeking to start a small airline. In reality their strategy is in the answer to a question every potential investor and potential banker will ask, 'What is the long-term objective of starting this airline and how will it support itself?' The answer might be 'The long-term objective is to establish an airline to provide a combined passenger and cargo service directly between provincial towns. At present no other airline provides this service and our calculations show that we can provide the service and achieve a profit.' This is a statement of strategy.
- Later in the book there is mention of an airline having an Audit Committee. One of the principal objectives of this committee is to ensure the airline's accounting records are accurate and complete and that its accounting policies are appropriate and up to date. Accurate accounts and reports are as important to the smallest company as they are to the largest. In a large well-established airline a committee may be needed because its operations are complex, but in a smaller airline the objective can be met by a director meeting the airline's auditors once the annual audit has been completed, to get their views on the accuracy and appropriateness of the airline's accounting records and principles. The director can then report back to the full Board of Directors. If this approach is used, the nominated director should not be the Chief Financial Officer (CFO) because there is a conflict of interest and the value of an independent view from the auditor may be lost. Ideally the director should be an 'independent non-executive director' (i.e., a director without any managerial function and no financial connection to the company other than receiving directors' fees).

As the airline develops and grows, the solutions in the examples above change or need to be changed. At some stage the definition of the airline's strategy may need to be expanded. Similarly, when the size justifies it, the airline may move towards forming an Audit Committee. The governance practices need to stay under review as the airline and its circumstances change.

Understanding what good governance is and using the principles sensibly in ways which are appropriate to the size and circumstances of the airline will help the airline in discussions with a range of people and institutions such as:

- Potential investors, who will want to understand what the airline seeks to achieve and how it plans to achieve the objective. They will also want to assess the management's ability and probability of achieving its strategy and achieving a return on any money they may invest.

- Bankers, financial institutions and leasing companies, who have a similar interest as the investors and in addition will need to assess the new airline's ability to meet its cash commitments and repay any funds advanced.
- Regulators, who in addition to having to establish whether the new airline will be able to meet the statutory requirements for a safe operation, are also interested in ensuring that the airline will be well managed and financially secure.
- Suppliers, including insurance companies, whose interest is in the potential size and development of the airline's operation and whether it will be well managed as well as assessing whether the airline is likely to remain solvent.
- Staff, who will invest their skills and careers in the company, and will want some assurance that the new airline has a good chance of being successful and generate sufficient cash flow to meet its costs. There may also be a second level of interest which is whether the new airline will be an organisation the staff will be proud to work for; good morale should not be under-valued.
- Revenue generators and cargo shippers, where their interest will be how well the airline will be run and whether it is likely to be able to continue to operate in what is a highly competitive business.
- Other stakeholders. It may seem unlikely that at this stage a new airline needs to consider the interests of others. If, however, the airline's plans will add to the noise level at the airports served and/or mean that flights will take off and land over a populated area and/or are likely to increase the traffic on the roads serving the airports in the airline's network, or for any other reason have an effect on others, interested individuals may coalesce into a group that needs to be communicated with and reassured.

With all of these potentially interested parties it is worthwhile for the board of a new airline to think about the various aspects of governance. Knowing how the airline will be run and whether it will be run well, will be reassuring not only to all the interested parties mentioned above, but to the original founders of the airline.

Compliance

The detail of corporate governance is full of lists, such as the examples in this chapter. There is a list of duties to be considered for the role of the Board of Directors, and a list of duties for individual directors. It can be easy for a Board of Directors to fall into the trap of apparently complying with governance guidelines on the surface but not following the intention of the guidelines. As an example, every company needs to hold an AGM which is a meeting between the company's directors and its shareholders. If the meeting is arranged in a location that is difficult to travel to and at a time which is inconvenient, the company will be 'ticking the box' that says they must have an AGM, but in practice have made it virtually impossible for shareholders to attend, hence are not following the spirit of the governance.

There are frequent comments that some companies take a 'tick the box approach' to implementing corporate governance, meaning that the company has taken a narrow, bureaucratic attitude towards the subject rather than discussing the principles, using them as part of general management and applying them to the company's circumstances. There are times when a checklist is not only useful, but essential such as, for example, assisting pilots when flying an aircraft. But those checklists are designed to ensure that things are done in the right order and nothing is forgotten. Using a 'tick the box approach' with all aspects of governance means doing something solely because it is on a recommended list and in doing so apparently satisfying some rule or regulation or best practice guide. But in reality, this just follows the appearance, not the content, of good governance. In the real world, Boards of Directors do need to comply with some lists (e.g., the documents and returns required to be sent to government departments), but these are more like a pilot's checklist.

Similar to the 'tick the box approach', there is the danger that some functions of the board are performed without enthusiasm or commitment or real interest. This might happen, for example, when the board meets to monitor the company's performance by examining financial or operating reports, or approving an operating and financial plan for the coming year. Directors may be tempted to assume that they do not need to understand the information or to question the result achieved or planned or to compare it with trends and competitors, but nevertheless approve the reports. This approach does not meet the requirement to energetically monitor the company's performance.

Earlier in this chapter when discussing the adoption of good governance in a small airline, there is an example of how a small company can comply with the recommendation to ensure its accounting records are accurate, complete and use the most appropriate accounting standards in a way other than having an Audit Committee. In this case the airline is complying with the spirit of good governance, but after thorough consideration, has decided that the objective can be fully met in a way other than through a committee and in a way which is more appropriate to its circumstances and its resources. The airline is following the substance of the recommended good governance, but not the usual form. In addition, it also reveals that the airline sufficiently understands what the guidance is seeking to achieve and is confident it is achieving the objective in a way that best suits the company at the time.

Particular governance problems for airlines

The nature of the airline business means that when an airline is establishing a system of corporate governance it may face some difficulties and challenges which are not found in all other industries. These include:

- Having a significant absentee workforce. Cockpit and cabin crew spend virtually all their time away from the airline's offices. This makes it difficult to communicate with them on plans and results and for managers and directors to answer questions from staff.
- Unpredictable business risks which can make organising finances to ensure continuing solvency a challenge.
- An airline's cost structure and the variations in revenues which make it difficult to make secondary capital investment decisions (e.g., re-configuring a cabin).

There is not a standard way to deal with these unusual problems in applying good governance principles. In the following chapters there will be discussion of these types of difficulties.

The key

Every company and every airline can benefit from having good governance practices which are kept under review and revised when there is a better way to achieve the objective. Adopting good governance practices should begin on the first day the airline is founded and those practices reviewed and developed as the airline changes. Good governance will support an airline and is part of developing a good reputation with its shareholders, financiers and stakeholders. The key to embedding good governance into the organisation is for the directors to fully understand the objectives and usefulness of the various recommendations and how they will help the airline not just in good times, but also when there are difficulties. A frequent comment when a major problem develops with a company is that some aspect of governance has been badly handled or is missing from the way the company really operates. A very important point is that once the governance practice has been adopted by the airline it should be the way things are genuinely done and not just be an empty formality. Good governance does not guarantee success, but it does help achieve it.

2 The Board of Directors and how it works

Organisation

The days when a company's Board of Directors was a group of people who knew each other well, who met periodically to chat about what was going on in the company, followed by a good lunch and then forgot about the company until the next meeting, have long gone.

Being a company director is a professional, serious, job with responsibilities, performance standards, guidelines, assessments and potential liabilities. There are recommendations on how much time a director should be able to devote to a company's affairs and how many directorships an individual should have. The operation of the board needs to be well organised so that the range of skills and experience of the board members are well used and their discussions are focused on the company's important issues. This is really not unreasonable when considering how important it is that each company performs well. The board is responsible for working in the best interests of the company and achieving its success. Even if the board conclude that the desired success cannot be achieved, it is responsible for plotting a path which will minimise any loss to creditors and shareholders.

The Board of Directors exists to make decisions which will help ensure the company's success. The board's decisions will not always be completely correct, but nevertheless they are generally made in the sincere belief that the outcome will be successful for the company.

Types of directors

In most countries the laws covering the operation of companies treat all directors as being the equal, regardless of their title or specific responsibilities, but in business, although all directors have the same overall responsibilities, there are different types of directors and it is important to understand the differences. Just as there is no generally accepted single definition of the roles and duties of a director, there is not one single definition of a 'director'. Rather than discuss all the options, this book will use a working definition for a director as being: 'A person, working with others, jointly to accomplish the long-term success of a company, by setting goals, monitoring actual performance and setting ethical standards.'

Whether a person is a director is defined by what they do and their responsibilities, not their title. It is not unusual for a company to give individual senior mangers the title of 'director' even though the person is not a member of the board and does not have the responsibilities of a director. Titles such as 'director', 'deputy director', 'local director' or 'divisional director' are quite common. There is the risk that using these titles will lead to confusion and that individuals dealing with a faux director may believe they are dealing with a member of the company's board. It must be made clear to anyone being given the title, but not the responsibilities, of director, that they must not give the impression to third parties that they are on the Board of Directors. The various 'director-type' titles are normally only given to trusted senior managers, so in most cases the risk of misrepresentation is probably small, but it must be remembered that risks exist. There is a danger for the individual title holder, that they may be held liable for some error or omission with the directors on the board. In most jurisdictions a court will decide on the responsibility for a board decision based on the input each director made. Thus, a non-board director may not be involved, but if they advised on a matter, the risk does exist. The status of all directors is usually shown in the company's annual report. Frequently directors who are not on the board are described as 'executive officers' or some similar title.

The main two types of director are:

- 'Executive director', who is a member of the Board of Directors and who also has executive/management responsibilities within the company.
- 'Non-executive director', who is a member of the Board of Directors but who does not have any executive/management responsibilities within the company.

Ideally each executive director should have a contract with the company which defines their responsibilities as an executive of the company. One of the executive directors will probably be the leader of the company's management team with a title of 'Managing Director' or 'Chief Executive Officer' (CEO) or similar. At the board meetings executive directors can give advice to the board on the operations of the areas they manage, but when decisions are made they should be influenced by what is in the best interests of the company as a whole, and not necessarily the area they manage. For some executive directors this sort of conflict can be difficult to handle. Once the board has made a decision, all directors must publicly support it regardless of their personal view or the effect on an area they manage, and this is often called 'cabinet responsibility'.

There are generally two types of non-executive director:

- A non-executive director, is an individual director on the board, who does not have executive/management responsibilities within the company, but may have some other financial connection with the company. For example,

they may be employed by a company providing accounting services to the company or may be a director who has been nominated to the board by a substantial shareholder.
- An independent non-executive director ('INED'), is the same as a non-executive director except that they do not have any financial connection with the company other than receiving directors' fees and they do not have any association with a particular shareholder.

Many jurisdictions have guidelines on what is and is not a financial connection with a company, but there is not a universal standard as yet. Every board will benefit from having independent directors because they are able to give an objective view on proposals, plans and results, and will bring experience of other companies and, possibly, different industries. Access to independent views can be an antidote to the danger of having an inward-looking and inward-thinking board. A small company will benefit greatly from having at least one, and preferably more, independent director on its board. A start-up airline will particularly benefit by plugging into the experiences of independent directors.

Quite often in a 'joint-venture company', which is a company formed by two or more individuals or companies to operate a business, there is a requirement that board decisions are to be made by a majority of directors, with a minimum number of the directors nominated by each shareholder voting. In order to ensure the minimum number of directors required for decision-making are always able to attend board meetings, it is not unusual for each director to appoint an 'alternate director' to stand in for them should they be unavailable to attend a board meeting. The alternate directors are empowered to perform all the duties of the director at the meetings the director cannot attend. For this arrangement to work efficiently each alternate director needs to stay up to date with the affairs of the company and this is the responsibility of the director appointing the alternate.

There is yet another type of director; a 'shadow or de facto director', who is someone who, although not appointed as a director of the company, frequently and routinely gives instructions to the board of the company, which the board usually acts upon. This situation can arise when a specialist manager, for example, an in-house lawyer or financial manager or engineering manager, routinely gives instructions relating to the company's operations to the board. The sort of actions which may cause someone to be deemed a shadow director are, taking complete responsibility for the financial affairs of the company or routinely negotiating on behalf of the Board of Directors. A manager may be acting as a director without realising it. The consequences of being a shadow director include the individual assuming all the responsibilities, duties and liabilities of being a director, while not being covered by the directors' liability insurance. There is a more sinister shadow director, who is someone who habitually instructs individual directors on the board on how to vote on issues and what decisions to make. In this case the shadow director assumes all the liabilities of a correctly appointed director and in addition, in some jurisdictions, there may be criminal consequences.

Chair

One amongst the directors will be appointed to manage the board. There is a range of titles for this position including 'Chair', 'Chairman', 'Chairperson'. Regardless of the title used they have the responsibility of managing the business of the board by:

- Ensuring the board deals with the development and implementation of the company's strategy.
- Acting as a link between the board and the company's management.
- Assisting in the introduction of new directors.
- Managing the discussions and decision-making of the board.
- Being a spokesman for the company.
- Being the Chair for all the meeting with shareholders, for example, at the company's AGM.

The Annual General Meeting is held once a year when the Board of Directors meet the company's shareholders to explain the results achieved in the past year, discuss the future of the company and the shareholders elect the directors for the following year. The minimum agenda items are usually listed in the jurisdiction's company law. The Board of Directors may meet the shareholders at other times during the year, but in most countries the legal requirement is for an annual meeting.

The Chair can be appointed from any of the directors on the board, executive or non-executive, but there is frequently a preference to appoint an independent non-executive director because of their element of independence.

The Chair of the board is not the chief executive of the company unless the two roles have been combined. It is quite common in the USA for the Chair to also be the CEO. In most jurisdictions there is not a legal requirement that the job of the Chair and the CEO are separate, but there is usually a recommendation that the roles be held by different people. The line of argument in favour of segregating the roles is as follows:

- One person running both the Board of Directors and the management of the company concentrates too much power and influence in one person.
- There is a risk that during an operating emergency a Chair/CEO will concentrate on handling and monitoring the detail of the emergency to the exclusion of thinking about the possible long-term implications and communicating with others.
- An important function of the board is to monitor the management of the company and this is best done by the independent directors, whereas

having a combined Chair/CEO could reduce the board's ability to oversee management.

It is clear from the number of companies which have a combined role that many companies believe that the advantages exceed the possible downsides mentioned above. It nevertheless remains the general recommendation that the roles are split.

One of the concerns with a unitary board is that one strong personality may unduly influence other directors. This situation can arise from the potentially difficult situation where executive directors must give independent views, but they also report to the CEO on management issues. This problem is largely counter-balanced by having INEDs on the board and an independent Chair. A similar situation can also arise on a board with a dominant Chair and a counter to this is to appoint one of the INEDs as the 'senior independent director' (SID) to act as a channel for other directors to voice any concerns when they feel they are not being fully considered at board meetings. A SID can also act as a confidant and advisor to the Chair and deal with the matters relating to the board Chair; for example, performing the Chair's performance assessment and chairing the committee to find and recruit a new Chair. This solution to a potential problem has not been universally adopted. When a SID is appointed it is usually on the board of a publicly quoted company and one of the benefits assumed is that having a SID gives shareholders an alternative channel for asking questions of the board. The arguments against appointing a SID are that the appointment risks splitting the board into groups and/or reduces the influence of other INEDs.

Others at the board meeting

Directors are not the only individuals who may be invited to attend board meetings. The Company Secretary will always attend and there may well also be advisors and managers from the company. Each of these individuals will be there for a particular reason.

Company Secretary

The Company Secretary is a senior officer of the company, possibly the senior legal officer, but is not a director of the company. The status and span of responsibilities for a Company Secretary varies from country to country, therefore it is difficult to produce a general list of responsibilities. At the most basic level, the duties of a Company Secretary are to ensure the company complies with the local Companies Act by keeping the required records and submitting the required reports to government departments. These include reports like a list of the company's directors and their declarations of interest, as well as keeping records of matters like shareholders' details and legal charges on the company's assets. In addition, the Company Secretary will keep a record of the proceedings at all meetings of the Board of Directors and meetings with shareholders. All of the 'required' items will be specified in the local Companies Act. Where there is not a requirement for a company to have a Company Secretary, the Board of Directors will be jointly responsible for submitting the required reports and keeping the required records. Although the Company Secretary is an independent function in most companies, in some jurisdictions smaller companies are permitted to appoint a director as Company Secretary so, for example, the CFO may also act as the Company Secretary. Where the responsibilities allocated to the Company Secretary are purely administrative and routine, they may be sub-contracted to a third-party company which specialises in this type of work or the Company Secretary may be a part-time position. Additional duties related to corporate governance may well be allocated to the Company Secretary, indeed in some jurisdictions the responsibilities are such that the position is titled the 'Chief Governance Officer' (CGO).

Where there is a newly formed airline it is likely that initially it will employ a professional firm to provide it with all company secretarial services it needs because expecting the directors to know all the details of the required legal reporting may be a burden for them. Once the airline becomes established and can afford the cost of a qualified Company Secretary, one should be employed and given the responsibility, in addition to complying with the legal requirements of the local Companies Act, of assisting in the development of good corporate governance within the airline. Once this stage is reached the Company Secretary position can provide whatever level of support the board requires.

Advisors

Some company Boards of Directors appoint a permanent advisor to the board on specialist subjects, for example to advise them when they are considering such strategic issues as entering a new market or developing a new product. The advisor is not a director and cannot vote on any decision, and is only there to advise on specific subjects. If the advisor attends every board meeting and/or discusses other matters, there is a danger they will be considered a shadow director, but this risk can be reduced if the advisor's appointment letter clearly limits the advisor to giving non-binding advice. There are boards which have a permanent 'Advisory Board' or 'Advisory Panel', and these are a group of experts who give advice to the board upon request. The members of the panel are not directors of the company and therefore do not have the liabilities of being a director. To avoid confusion, it is better to use the term Advisory Panel, but many companies like the implied status of Advisory Board.

An Advisory Panel will usually include a range of expertise from legal through finance and marketing to technical. There is a great deal of flexibility in the composition of the panel and how the advice is given to the company's board. The full panel may be asked to give combined advice on a particular topic or to individually give advice to the board. It is the flexibility and access to a broad range of expertise which makes the Advisory Panel approach such an attractive option. Frequently the remuneration of members of an Advisory Panel is less than that of a director because the risks and liabilities are smaller. The attraction to the members of the panel is they can give advice without the liability for making the decision. The board is not bound to take the advice of the panel or any individual on the panel.

Although it is generally the large international companies which form and use an Advisory Panel, this option should have substantial attractions for companies of any size, particularly newly formed ones. If a newly formed airline appoints a panel of advisors it could have access to expertise, both academic and practical, in areas such as operations, marketing, finance, information technology and engineering, at a relatively small cost. In addition, the panel could include expertise in topics which arise irregularly and are outside the expertise and experience normally found on an airline's board, for example, purchase and development of property. At the very least the new airline's board would have access to a team of experienced experts, who understand the airline's objectives, to advise and comment on the board's ideas and the airline's possible developments without the formality of appointing directors or retaining consultants.

Staff

Periodically, members of staff will be invited to attend board meetings, usually to explain and answer questions on a particular project or decision the board have been asked to make. The staff should be competent to give the board the information it needs, but the status and title of the staff are not important. The staff should only attend that part of the board meeting which deals with the particular project or required decision they are giving information on and then leave the meeting before a decision is made. The staff should not take any part in the decision-making process. Not only is this a very useful way for the board to get information and understand the implications of their decision, but it also gives the staff an insight into the information the board needs before it can make a decision.

Achieving a 'balance on the board'

The central purpose of the Board of Directors is to make decisions and policies that will guide the company towards achieving the long-term goals set and agreed by the board. The assumption underlying the formation and organisation of the Board of Directors is that the collective wisdom of the directors, jointly reviewing, discussing and deciding on courses of action will be more beneficial, constructive and useful than an individual director acting alone. There are instances where Boards of Directors have made a decision which did not have the desired result and it is clear that making decisions jointly is not a flawless process, but on balance the arguments fall in favour of the idea that better decisions are made by a group rather than an individual.

To maximise the chances of making decisions which will benefit the company, there needs to be a balance of skills and experience on the board; sometimes this balance is called 'board diversity' (this now frequently includes gender issues). In this context 'skills' are knowledge gained by training and study, while 'experience' is knowledge gained through the actual performance of skills. It should also be borne in mind that someone with experience in an industry usually also has contacts within that industry and with service-providers to the industry. These contacts may assist a company to get up-to-date advice on various topics and can be very helpful to a newly formed company, particularly a company like an airline which needs the support of many service-providers and regulators. Achieving the appropriate balance of skills, experience and contacts is the joint responsibility of the Chair of the board supported by a committee of the board usually called the 'Nomination Committee' or the 'Nominating Committee'. There will be more on this committee later in this chapter.

The starting point for deciding who to invite to become a director of a company is to decide just what skills and experience a company's board needs. The list of the desired requirements will change over time as the company grows, adds new operations or eliminates existing operations to concentrate on a core business. For example, the board of a small-scale newly formed airline will probably focus mostly on safety, operating efficiency, marketing, cash flow management and accurate reporting of results whereas a mature airline will add to this list such items as information technology and brand management. It is not sensible to change the composition of a board too frequently because directors, especially independent directors, need time to understand the company's long-term goals, the circumstances of the industry, methods of operating and its reporting systems. But change is needed periodically and should be planned for, as even a rose bush needs to be pruned periodically.

Deciding on the skills and experience looked for is something of a juggling act because there are many factors to consider, one of which is that to be effective the board needs to be of a workable size. A rough guide to the number of directors on the boards of airlines range from six to twelve, although there are exceptions at both ends of the scale. It may be tempting for a newly formed small-scale airline to have a small board made up entirely of executive directors, but this approach ignores the contribution that an independent director can bring to discussions and decision-making as well as the general recommendation is that every board will benefit from having at least one or two INEDs.

The desired size of the board can make it difficult to pack into the board all the knowledge wanted and this is a situation where an Advisory Panel can help. One factor the Chair needs to consider when thinking about the size of the board is that if there are an odd number of directors, including the Chair, it should be possible to avoid having a tied vote, assuming that no director abstains from voting. The factors likely to be considered when deciding on the desired composition or balance on the board are:

- Executive directors versus non-executive directors: one of the difficulties of a board which is composed only, or which has an overwhelming majority, of executive directors is that it can be difficult for the board to objectively monitor and discuss the operating results of the company. Some best practice corporate governance guidelines recommend that at least 50% of the board should be INEDs.
- Experience with the company: there is frequently a debate about the need for directors to understand how the company works and to have experience of the good and difficult times experienced by the industry, while on the other hand there is also the need to understand and evaluate current ideas to establish whether the company can benefit from adopting new approaches to operations. Airlines are classically an industry where it is beneficial to have years of experience, but also to understand the value of assessing potential developments. When thinking about the experience a director has on the board, it should be remembered that in some 'best practice corporate governance guidelines' a director may cease to be considered independent after serving on the board for six years or more. There is a growing trend for companies, including airlines, to include in their annual reports details of the composition of the Board of Directors showing the number of each type of director, their spread of experience in the industry and in other businesses. The figures for two, anonymous, airlines are presented in Table 2.1.

Table 2.1 Information on directors

	Airline 1	Airline 2
Years of service (%)		
5 or less	33	65
6–10	20	30
More than 10	47	5
Directors (%)		
Executive	13	28
Non-executive	40	44
Independent non-executive	47	28

Note: The percentages have been rounded.

- Specialist experience versus general business experience: there may be a tendency for the Chair to assume that the company needs to have many specialists (e.g., lawyers, bankers, engineers, accountants) on the board. Whilst these experts will all make a contribution it should be remembered there are also benefits from having non-specialist business men and women on the board as well, whether with airline related experience or with experience in unrelated industries. The Chair may, for example, want to consider whether a director with experience of the fast-food business could add value to an airline when considering customer service, queueing, pre-selection of meals and response to changing food fashions. There is a view that decision-making committees make better decisions when the members are from different backgrounds and experience. Also, that proposals are subjected to greater scrutiny when there are directors with different perspectives.

- Special considerations: it may be that some companies operating internationally (this would include many airlines) can benefit from having a director(s) who understands the different cultures served by the airline. It may be that this aspect can be, or is better covered, by including the expertise in an Advisory Panel. Similarly having access to customer groups may produce useful information, but this also can be achieved by adding to an Advisory Panel.

Achieving a good balance on the board is not just a matter of numbers and ratios or taking a 'tick the box approach'; the board needs to work together as a team to make decisions which will achieve the company's long-term goals. For this to happen, it is necessary to balance the personalities on the board and to create an atmosphere where directors can speak freely and challenge proposals without causing resentment amongst other directors. There is a real skill in producing a board which has the right balance of skills and experience within a size that is effective for making decisions.

Board committees

The span of responsibilities of the Board of Directors is quite wide. In addition, there may be areas of the business that require specialist knowledge or closer monitoring; for example, the importance and complexity of safety and the management of risk to an airline. Also, there is the need to closely monitor the airline's financial reporting and ensuring that good accounting standards are followed. To effectively deal with the range of responsibilities as well as to cover the specialist areas yet still keep the responsibilities within the board, most boards form committees of the board ('board committees') to do the detailed work while keeping the board fully advised by issuing reports or copying the minutes of their meetings to the board. The board committees should operate by making researched and considered recommendations to the board who, after discussion, approve, modify or reject each recommendation. This approach acknowledges the board's continuing responsibilities because delegating the detailed work to a committee does not mean that the board has delegated the responsibility for making a decision. Having the board committees make recommendations to the full Board of Directors for a decision ensures that all directors have a chance to examine the recommendation and that the board's 'cabinet responsibility' is maintained.

A company is not required to have board committees, but the majority of boards find them to be an efficient way to spread the workload of the board and deal with matters which are important to the company. If the company's shares are listed on a stock exchange, it is very likely that the stock exchange's rules will require that the board form certain committees and these are likely to be:

- Audit Committee responsible for monitoring financial reporting. Sometimes there is a requirement that this committee is also to deal with risk and its management.
- Remuneration Committee dealing with the remuneration of executives.
- Nomination Committee responsible for identifying successors to senior managers and directors.

There is not any limit to the number of committees a board can have, but there is general guidance that each committee should have two or three directors and that each director should not be a member of more than two committees. Individuals who are not directors of the company may be committee members and appointing additional specialist members can enhance the work of a committee, but non-directors should not vote on any recommendations sent to the board for approval or any other matter where a vote is required.

Each committee must have terms of reference which are agreed by the board. These terms of reference should be related to the needs of the company and its board, but when deciding on the terms of reference it is sensible for the board to consider the recommendations of good governance guidelines as well, so the board plugs into the knowledge of others. The terms of reference should be reviewed periodically – often this can be every three years, to ensure they are still appropriate. It is also sensible to review the operation and the membership of each committee regularly.

What board committees an airline should have

There is not any limit on the number of board committees a company or an airline can have, the number and their exact terms of reference will depend on the way the board wants to organise its operation and to some extent, the number of directors who are available to serve on the committees. An exception is that generally an airline's operating certificate will specify that the airline must have a flight safety section or department which reports directly to the senior operating director.

The following section deals with the main board committees which an airline is likely to find useful: Audit Committee; Finance Committee; and Remuneration Committee. There may be other board committees for such matters as the environment or risk, and in addition there may be company committees which also deal with other important matters but they are not always board committees. Examples of the latter are:

- Management Committee, responsible for monitoring the day-to-day operations of the airline.
- Safety Committee, overseeing and advising on all flight and operating safety issues. The operation of this committee must comply with the requirements of the airline's regulatory authorities which will often specify how flight safety is to be monitored and managed.
- Staff Committee, charged with being a forum to discuss matters relating to staff; not just remuneration, but training, systems for dealing with staff and the two-way communication between the board, management and staff.

There are potentially many more useful committees that may be functioning to effectively operate the airline, but which are not board committees.

Audit Committee

Although having an Audit Committee is not a legal requirement in every country it is, however, recommended as part of good corporate governance, which concerns the overall management of a company. If it is a legal requirement, the law will define the committee's duties. If the airline's shares are quoted on a stock exchange (i.e., it is a 'public company'), it is almost certain that a stock exchange listing the airline's shares will require the company to have an Audit Committee and will list its minimum responsibilities. The Board of Directors can allocate duties to this committee in addition to any legal responsibilities or any imposition by a stock exchange. The usual duties of the Audit Committee are to:

- Establish that the company's accounting policies are appropriate.
- Ensure the company's financial reporting is complete, accurate and complies with current 'Reporting Standards'; these are national (or international) accounting and reporting standards applicable to all companies.
- Monitor the airline's 'internal controls' which are the systems designed to prevent fraud and ensure accurate and timely recording of transactions.
- Ensure compliance with the airline's accounting and management policies.
- Monitor the ways the airline's risks are managed. If there is a committee which is responsible for risk management this responsibility will be passed to that other committee.
- Choose and monitor the performance of the company's 'external auditors' who are independent accountants employed to review and confirm the accuracy of the airline's 'statutory reports' (i.e., those reports required by law).
- Monitor the performance of the airline's 'Internal Audit Department'; this is a separate department responsible for checking the accuracy and efficiency of the company's systems. In some structures the Internal Audit Department will report to and support the operation of a Risk Management Committee.

The members of the Audit Committee should all be INEDs with some experienced in finance or accounting.

Finance Committee

Finance Committees are not currently required by law, but most companies, including airlines, establish one in one form or another, to help manage the financial position. Its duties are allocated and agreed by the Board of Directors. Hence there can be a variation company by company. In general, the duties for an airline's Finance Committee include:

- Monitoring the company's general financial position and 'capital structure'; (i.e., the company's equity and borrowing position).
- Recommending the timing and sources of fundraising.
- Reviewing 'financial exposures'.
- Reviewing the reporting of the airline's financial position.
- Monitoring financial relationships with third parties; these are entities to which the company owes money (loans), or which owe the company money (deposits), together with the amounts owed.

Membership of the committee usually consists of some executive directors and INEDs together with staff from the airline's finance and/or treasury functions.

Remuneration Committee

The essential principle underlying the operation of this sub-committee is that no director or senior executive should set, or be able to influence, their own remuneration package. Regardless of the need to avoid excessive remuneration packages, it is important that individual packages are fair, and are seen to be fair, between the seniors and executives. In addition, there has been considerable concern expressed by the shareholders of some large and publicly quoted companies as to the construction and size of the remuneration packages for some senior management as well as directors. Frequently the packages have been seen as being disproportionate to the company's result.

The main duties of this committee will include:

- Monitoring the remuneration packages of senior executives in the industry and for airlines of a similar size and operations.
- Monitoring the process of setting the remuneration for senior staff and staff in general to ensure that, as far as possible, it is fair and unbiased.
- Recommending remuneration packages for individual senior executives.
- If targets are included in remuneration packages as a trigger for payments, ensuring that those targets are reasonable.
- Recommending the fees and expenses for directors. Most companies pay a fixed annual fee with additional fees for membership of a board committee,

but there are some who pay fees only for the meetings which a director attends.

Because of the nature of the responsibility of this sub-committee, all its members should be INEDs.

Nomination Committee

The role of this committee is to identify potential candidates for appointment to the Board of Directors and to the senior management of the company. To do this the committee needs to understand the skills and experience needed by the board and have access to the assessments of the board's operation as a whole as well as for individual directors. The processes need to be systematic and transparent. Usually the Company Secretary is the secretary of the committee. The committee's terms of reference will include:

- Regularly reviewing the size and composition of the board, recommending changes if appropriate.
- Reviewing the succession plans for directors and senior executives.
- Understanding the available pool of potential candidates and identifying sources of potential candidates for appointment to the board or senior management.
- Reviewing the results of the regular assessments of board performance and the assessments for individual directors.
- Reviewing the composition, performance and succession plans of all other board committees.
- Preparing and overseeing an induction plan for new directors.
- Staying up to date with appropriate legislation and best practice.

Usually the committee will have a minimum of three members, all of whom will be INEDs.

Risk Committee

Many companies' Boards of Directors, and airline boards in particular, have been considering, or re-considering, how they deal with the risks the companies face. There has been a realisation in aviation that in addition to operating safely in the air and on the ground, there are other serious risks to an airline's results, reputation and brand image. This is clear from the examples of the effects on airlines of the theft of customers' data, military conflict in other countries, political disputes, civil unrest, global pandemics and airport and government charges. These all mean that the definition of risk has been broadened. To deal with this development airline boards are re-thinking how to identify and cope with this wide array of risks. In general, the thinking is that the traditional groups responsible for dealing with specific risks should remain in place. These usually are, flight safety with the Safety Committee, financial risks with the Finance Committee, operating safety and risks with an Operations Committee, but that other more general risks should be identified and monitored by a separate group reporting to a board committee, the Risk Committee. This approach has the advantage of keeping the identification and management of safety and risk as close to actual operations as possible.

The terms of reference for the Risk Committee, which will be separate from any Safety Committees, will generally include:

- Identifying and defining the types of risk the airline faces: strategic, regulatory, financial and business and operations.
- Liaising with the airline's other risk management groups (e.g., Safety Committee) to ensure all risks within their responsibility have been identified.
- Producing recommendations for the board on how to manage risks not covered by existing groups.
- Monitoring the operations and deliberations of all the airline's risk management groups.
- Reviewing the airline's long-term plans to identify any new risks which may be faced (e.g., when buying a new aircraft type or serving a new destination).

This committee will concentrate its efforts on identifying and managing the business risks involved in the airline's operations, monitoring the systems already in place to manage business risk, as well as identifying new risks. It may be that as a result of the monitoring process the committee may conclude that the chance of a potential risk seriously affecting the airline has reduced to a bearable level.

The committee will frequently consist of two INEDs and some executive directors and executives involved in each risk area (i.e., Safety Committee, Finance Committee and Operations Committee). It is likely that the airline's internal audit function will support the operation of the Risk Committee; it rather depends on whether or not the internal audit function focuses only on financial reporting and systems, in which case it will work with the Audit Committee or has a responsibility for business and operating matters as well. Then, the Risk Management Committee will most likely be its home.

Governance Committee

Good governance is not just an agreed set of policies and practices recorded in a manual; it is a record of how the business is actually being guided and operated. There is a danger that agreed governance policies are circulated to managers and operating staff, but after a while these are forgotten or modified without record or approval. It is important that the company's policies and practices are kept under review and there is some monitoring of whether the current policies and practices are still appropriate and are being followed. There is always a danger that anything reduced to a manual is simply put on a shelf and forgotten.

There is a growing realisation that it is necessary to monitor the effectiveness of the company's policies and practices and the degree to which they are being followed. Frequently this is done by the board establishing a Governance Committee and using the services of the company's Internal Audit Department to monitor and report on compliance. This approach is not yet a general practice and is currently most frequently seen in banks, professional companies, education institutes and public utilities such as water companies. If an airline's board decides it wants to ensure that its governance policies and practices are up to the current best standards, relevant to the airline's current operations and are being followed it can form a Governance Committee. The role of the committee is to be responsible for assessing all aspects of the airline's corporate governance policies and practices.

The terms of reference of the committee are likely to be:

- At least annually to review and report to the Chair of the board on the effectiveness of the airline's current governance policies and practices and to highlight deviations from them recommending possible improvements and changes.
- Each year to review the role, composition and operation of the airline's Board of Directors, including any board committees, taking into account the basis and method used by the board to assesses its own performance, and recommending possible changes or improvements. The board's self-assessment process might be seen as the board marking its own examination paper and for this reason, in some jurisdictions, large public companies are required every few years to have the board's performance assessed independently by a commercial company, professional advisors or a team from the local Institute of Directors.
- Regularly review the terms of reference, frequency of meeting and efficiency of operation of each board committee and report to the board on whether any change would be beneficial.
- Regularly review any board policies on matters involving directors such as avoidance of conflict of interest and policies related to conduct and ethics and recommend any changes to the board.
- Periodically review the frequency, format and quality of the information routinely presented to the board, and also consider whether additional information would assist the board's operation, recommending any beneficial changes. This work should be co-ordinated with the Audit and Finance Committees.
- Review the content and quality of the induction programme for new directors.
- Review any information relating in the airline's annual report relating to its system of governance.

- Recommend and develop additional governance policies and practices which will assist the airline to achieve its objectives.
- Monitor the general development, advice and recommended best practices of governance for companies and boards of directors.

Committees and small companies

It is possible to have the view that board committees can only operate in larger companies and that for a small company, like an airline just starting its operations, establishing board committees is an unnecessary burden which at best is costly and at the worst is a diversion from the important job of establishing the new operation. This attitude falls into the trap of confusing the appearance or form of good governance with the substance of good governance. Committees are formed to help the board do its job but not act as corporate status symbols. Understanding the contribution which each committee can make to the operation and development of the company will permit the board to look at other ways of achieving the same benefits as forming a committee.

Possible ways a newly formed small operation may get the same benefits with a simpler approach are with:

- Audit Committee: a small company or airline will be just as concerned as a large company to ensure that its financial accounting and reporting is accurate and of a high standard, but it may not have sufficient directors experienced in financial reporting to form an Audit Committee. The committee's role can be taken by an INED meeting the company's external auditors and reviewing and discussing the company's current accounting policies and practices with them. In addition, the external auditors can be asked to give the board a 'management letter' in which the external auditors comment on the efficiency of the company's accounting processes. The INED will then report to the full Board of Directors. As the company or airline grows, a point will be reached when a formal Audit Committee can be justified. In general, it is preferable for the company to form its Audit Committee as soon as is practicable.
- Finance Committee: it is not so easy for a small airline to replicate the function of a Finance Committee on a small scale. Certainly, there are banks and financial institutions who are capable and willing to give advice, though the difficulty is to ensure that the advice is not biased in favour of the bank's products. A small committee chaired by an INED with the CFO and a representative from a minimum of two banks can provide financial advice to the board. With this approach an important factor is to ensure that the airline has accurate records of its financial needs and exposures.
- Remuneration Committee: even the owner of a one-man operation needs to be sure the one employee is remunerated fairly in line with the market. In a smaller airline the responsibilities of this committee can again be taken on by an INED who will be responsible for staying in contact with

other airlines, monitoring information published by trade associations and job advertisements also maintaining contact with recruitment specialists. With this background the INED should be able to advise the board on remuneration.
- Nomination Committee: every employee in an organisation is important and the smaller the organisation the more significant it will be if a staff member leaves without a replacement. A solution may be for an INED to be responsible, once a year, for reviewing the retirement plans and performance reviews for each employee.
- Risk Committee: the airline's CEO, an INED and the airline's safety manager(s) could meet regularly with the airline's insurance brokers to discuss and get updates on risks and their management.
- Governance Committee: it would be helpful if all directors, or at least all the INEDs, on the airline's board become members of the local Institute of Directors and participate in its courses passing the information regularly to all the members of the board. The cost of this should be borne by the airline.

Not all the solutions to the business needs and functions of board committees are wholly satisfactory, but they go some way to bridging the gap until the airline is of sufficient size to perform the functions in a more formal way.

Effectively running the Board of Directors of even a small company or airline involves a lot of processes and many decisions. At every turn there is the need to balance talents and competing needs to meet expectations but to keep the administration in proportion to the size of the company's operation.

3   Organising the board

The rules

Being a director of any company, however large or small, is a real, professional job and most countries have some form of qualification process in directorship and/or corporate governance which can end in membership of a professional association. Part of the development of professional standards for directors has been for various bodies to discuss and give guidance on a director's responsibilities and the rules and guidelines that are involved. There are also rules and recommendation on how the board should operate most effectively and how meetings with shareholders should be run. The rules and guidance come from a number of sources:

- Company law: in most jurisdictions the law gives a list of a director's duties and responsibilities and these tend to be rather general statements. The difficulties of producing an absolutely complete list have already been discussed. In addition, the local law governing companies will list the documents and records the company needs to keep as well as the reports the board must make to government authorities. Inevitably it will also list the penalties for failure to comply with the law.
- The company's constitution: traditionally this comprised one document which included the company's Memorandum of Association and Articles of Association, frequently referred to as the company's 'Memo and Arts' or 'M and A' (not to be confused with Mergers and Acquisitions) or 'Constitution'. There is a trend to reduce the size and content of the Memorandum of Association to a simple statement of the objective of the company and include all the other information in the company's Articles of Association. The Memorandum of Association includes the basic information about the company, its name, the reason for forming the company (e.g., to operate an airline), the amount of its capital and the names of the individuals who formed the company, also called the 'subscribers'. The Articles of Association essentially describe the way the company is to be run, dealing with such matters as the responsibilities and powers of the directors, the process on how the board should arrange to vote on

decisions, how directors should be appointed and removed, and how share certificates should be issued. In other words, it is a general statement of how the company should be run at board level as well as the relationship between the Board of Directors and the company's shareholders. It is likely that the jurisdiction's company law will include a model set of Articles of Association which the company can adopt in total if it wishes or can, within limits, draft its own set of Articles based on the model.
- Government regulations: in most jurisdictions government departments have the right to require companies to report information to them regularly. For example, the Registrar of Companies will probably require an annual return which updates their records with the essential information of the company such as, the names of the current directors, Company Secretary and shareholders. There will probably be penalties if the returns are not received on time and some penalties will apply to each day the return is overdue.
- Stock exchange listing rules: if the company's shares are traded on a stock exchange it is likely the stock exchange will have their own reporting requirements. For example, the listing rules will state the minimum amount of information to be included in the company's reports to shareholders including the financial and operating information as well as information on 'ESG', which is a commentary on the company's environmental, social and governance matters. The listing rules may also impose an obligation on the company to include certain features in the company's corporate governance, for example, to have an Audit Committee. Failure to comply with the exchange's listing rules may result in the company being fined or some restriction being placed on the trading in the company's share. The company's reputation may also suffer because frequently the name of companies not complying with the listing rules are published.
- Aviation requirements: in addition to the rules, guidelines and best practice advice which apply to every company, an airline's operation will be licensed by a specialist government authority or department, frequently called something similar to 'Civil Aviation Department' or 'CAD'. This authority will have the power to license the airline to operate, but will also impose conditions and rules on the airline. The majority of the rules will relate to the safe operation of the airline, but are likely to also include matters which influence how the airline is organised. Two examples are as follows: first, it will probably be a requirement that the airline designate a suitably qualified individual to be responsible for flight safety in the airline and that the individual report directly to the airline's most senior executive director, perhaps the CEO; second, it is likely that the CAD will monitor the financial position of the airline to ensure it has sufficient funds to operate and to operate safely.
- Professional associations or institutes: within a jurisdiction there are often a number of professional bodies which have an involvement in a company's governance, hence a director's duties and responsibilities. In

most jurisdictions the local accounting association is likely to be responsible for issuing accounting standards, Financial Reporting Standards and increasingly, as the requirement of reporting standards expand to include operating and ESG matters, on reporting on the company's business as well. The Association of Company Secretaries will also probably issue its own guidelines and best practice guide on complying with all of the company's legal obligations as well as some guidance on governance. An important organisation involved in governance is the association or institute which represents the directors themselves, often called the 'Institute of Directors' ('IoD'). This organisation exists to help and represent directors. It usually does this by running training courses, issuing guidelines for directors and best practice advice and periodically issuing bulletins on issues and matters which interest directors. The IoD may also run courses which lead to some form of professional qualification for directors.

The regulations surrounding the operation and corporate governance of an airline are significant and the responsibility for the corporate governance falls onto the shoulders of the Board of Directors.

In practice, the responsibility for meeting legal routine reporting requirements will be assigned to the airline's Company Secretary who will be familiar with all the requirements. Each new director, whether experienced or not, should get a full briefing on the reporting requirements which apply to the company, usually from the Company Secretary. Part of the 'induction programme' (i.e., the programme which briefs each newly appointed director on the governance and operation of the company), will include a review of the details of all the regulations, from all sources, statutory as well as the company's own, as to what is required and the penalties for not complying. At each meeting of the airline's Board of Directors, the Company Secretary, will advise the directors if and when the implementation of a decision will require either a report to, or permission from, a regulatory authority. Allocating the responsibility to the Company Secretary does not mean that the board can ignore the legal and regulatory requirements, as they remain responsible for ensuring that all legal requirements are met, but allocating the responsibility to an individual ensures that the primary responsibility is clear.

Induction programme

The director's induction programme is important to every newly appointed director and the company. Even if the new director has served on the board of other companies, an induction programme is important and useful because companies have different organisations, internal rules and policies. The objective of the induction programme is to get the director up to speed with the company and how it operates. It will ensure the director understands not only how the airline works, but also the regulations influencing its operation. A good programme makes the director more productive sooner and in addition helps the director to quickly become a member of the board team.

A balanced induction programme really needs to be individually constructed for each director, and although each programme will have a majority of items which are common, there are some items which will have a different emphasis. For example, a professional accountant joining the board as an INED will probably be happy to examine the airline's budgets unaided, but might need help in understanding the obligations and detail of the airline's 'Air Operators Certificate' (frequently called an 'AOC'). A typical induction programme will include:

- Copy of the airline's Memorandum and Articles of Association.
- List of directors' duties and responsibilities together with guidelines on good corporate governance. This information is often provided by the local Institute of Directors.
- Copy of the stock exchange's listing rules if the airline's shares are publicly quoted and traded.
- Summary of the government's and other regulatory authorities' and stock exchange's reporting requirements, deadlines and penalties for non-performance. For an airline this summary should include the requirements in the airline's AOC.
- Airline's organisation chart with the names of the senior managers.
- List of the airline's subsidiaries and associates together with a list of the senior management of each operation.
- Last two years' annual reports including the financial section.
- The current year's approved budget and forecast for the year's result together with any long-term plans.
- Long-term plan (i.e., the airline's strategy).
- Board of Directors' work plan for the coming 12 months.
- List of the airline's advisors, including bankers, lawyers and auditors with the name of the contact in each organisation.
- List of the major shareholders.
- Minutes of the previous two Annual General Meetings and previous two years' board meetings as well as meetings of the board committees.
- Current papers on the airline's risk and risk management.

This is a lot of paperwork to go through, but it is important to get a complete picture of the airline's current position financially and operationally. In addition, once the information in the paperwork has been digested by the new director, it is important that they start to understand how the airline actually operates and this will involve making visits to key locations, operations and subsidiaries, and meeting the staff working there. At the end of the process the new director should discuss the information given during the induction programme with the Chair to clarify any issues in the director's mind and additionally discuss the current aviation industry's condition and any particular challenges for the airline. The final item in the programme is to discuss the director's needs for director education and 'CPD' (i.e., 'Continuing Professional Development') which is a continuing programme of education in the profession of being a director to stay up to date with legal and best practice guidance and also to increase the director's knowledge.

This is a formidable programme and workload for the new director, but it is important that the director becomes a contributor to board discussions and decisions as quickly as possible.

The contents of the programme for a newly appointed director promoted from within the airline will be much the same as for a new director from outside the airline. Probably the promoted director will not need to spend much time on the current organisation chart and the content and requirements of the AOC, but the other areas will require their full attention.

The contents of the induction programme will be much the same regardless of the size and stage of development of the airline. It is assumed that a small, start-up airline will be as well governed as a larger mature airline.

Board's annual work plan

The workload of the average Board of Directors is not normally spread evenly throughout the year, as there are peaks and troughs. Regular agenda items at each meeting will include such items as to:

- Review and agree the minutes of the last meeting.
- Declare conflicts of interest.
- Review and discuss the financial and operating performance for the past period(s) compared to the short-term plan.
- Review and discuss the report on the current funds position (i.e., its holdings of cash, deposits and bonds), any outstanding financial commitments (e.g., forward sales or purchases of currencies) together with the current financial position (i.e., a statement of how the company is being financed, for example, by equity and loans), all compared to the short-term plan.
- Review and discuss the forecast funds position in the company's home currency and in the major currencies used by the company, compared to the short-term plan.
- Review and discuss the forecast financial position.
- Review and discuss the minutes of board committees.

There are, however, items which only arise once or perhaps twice a year and tend to demand immediate board attention, including to:

- Review, discuss and approve the annual budget or profit plan including the cash forecast and forecast financial position.
- Review, discuss and approve the annual financial statements and annual report.

- Review, discuss and approve the plan for capital expenditure.
- Review customer satisfaction survey results.
- Review and examine the staff morale and feedback survey.

The third category is for matters which are important, but which can be dealt with at any time during the year, such as to:

- Review board performance.
- Compare the airline's financial and operating performance with competitors.
- Review company policies for the environment, staff, investor relations, etc.
- Review the current strategy, including a review of traffic rights, the timing of aircraft purchases and other major capital expenditure.

The art of producing an effective work plan for the board is to assess the workload involved in each item and try to produce a balanced plan which spreads the workload as evenly as possible over the number of board meetings in the year. Usually the draft work plan for the board is prepared by the Company Secretary in discussion with the Chair and CEO and shown to the full board for approval. Experience of how effective the work planning is will be built up over a few years and the work plan can be refined and made more efficient each year.

It is important that directors are allowed sufficient time to review and discuss each item on the agenda for each board meeting and still have some time available for 'Any Other Business', which are items worthy of discussion after all the matters on the board's agenda have been dealt with and finalised. 'Any Other Business' allows directors to raise issues that are important to them.

Virtually all of the items on the board's agenda for a large mature airline will appear on the agenda of a smaller airline, as the basics are the same. It may be that a smaller airline will not have board committees but rather a report from an INED. It could be that the time allocation between agenda items will vary between mature and new airlines, but the important consideration is that each item is discussed in appropriate detail, not that the agenda is completed within a certain time. Directors must be prepared for board meetings to continue for longer than anticipated if the directors feel they need more time to deal with all the agenda items.

Adopting a work plan which covers all the board's meetings for the following 12 months does have a potential risk. Directors may be tempted to focus only on those items in the work plan and believe that dealing only with these items will mean that everything of importance has been covered. It is easy for a board to get out of the habit of critically examining the probable future of the airline and its current performance from an independent viewpoint, but rather just dealing with routine matters. Perhaps they simply compare the airline's performance with the airline's own plan rather than with what is happening currently in the marketplace and with what others are achieving. This is the often mentioned 'tick the box approach'. There are ways that a board can try to avoid falling into this trap; which ones are used depends on the culture of the airline's board:

- Have a majority of INEDs on the board and limit their appointment to a short period, such as three years. Stagger the introduction on the new INED so there is always a new INED on the board. This may result in the operation of the board and the airline being reviewed frequently.
- Use the 'Any Other Business' part of the board meeting to encourage directors to compare the airline to others in the industry or companies in other industries, for example, by asking the questions 'Why can't passengers order their food in advance or when they check-in?' There may be operational reasons why this cannot or should not be done, but asking the question may prompt fresh thinking on the subject of onboard meals and customer service.
- Ensure all directors regularly receive aviation industry magazines and summary of economic briefings from banks and investment houses.
- Periodically have briefings from independent experts on aspects of the industry, to keep directors up to date with the thinking in the industry.

None of these are guaranteed to work all the time, but are an attempt to keep the board relevant, fresh and thinking about the airline and its future.

Conflict of interest

This is a subject which is very important for every director and carelessness in this area can cause difficulties for a director and the board. In Chapter 1, the list of a director's duties included, 'act honestly....', 'exercise their powers for the benefit of the company' and 'avoid conflicts of interest'. Each of the first two duties reminds a director to adhere to the third and to avoid conflicts of interest. In business a 'conflict of interest' is the situation where an individual, in this case a director, will, or is likely to, benefit from the result of a decision they have been invited to make. In other words, the director is not able to be involved in making a decision because they may benefit personally from it.

The possibility of having a conflict of interest does not bar an individual from being appointed a director and participating in the discussions and decision-making of the board, but if or when a director sees there is a conflict of interest or the potential for a conflict in a particular decision, the director(s) must take special care to avoid that conflict. It is not difficult to do and there are established and accepted practices, but to be safe the practices should be understood and followed. Sometimes it is not easy to decide whether a conflict of interest exists or might exist. An obvious case of conflict of interest is where an airline is considering awarding a contract to a contractor whose owner is a non-executive director of the airline. Another example where there is the potential for a conflict is where a non-executive director of the airline also provides a service to the airline either personally or as a professional in a company or partnership (e.g., a lawyer or architect). A less obvious case might be where the director of the airline is also the trustee of a charity which solicits donations from the airline. The potential for a conflict of interest does not bar an individual from joining a board; it can be argued that the definition of a conflict can be so wide that potentially almost every director will have one. The key is to handle any conflict or potential conflict openly.

The situation where a director of the company also provides a service to the company requires careful thought. A director should not get into the position where their duty to the company and their personal interests may conflict. A conflict of interest may be an actual conflict or a potential conflict or a perceived conflict. In most jurisdictions the law on conflict of interest is complicated and may not seem to cover all circumstances. In addition to the legal position, another important factor to be considered is whether there may be the perception of a conflict. A perceived conflict may damage the director's and the company's reputation. If it is suggested that a director, as an individual, should provide a service to the company e.g., to act as a property development consultant, for which the company will pay, the best and clearest option is either for the director to decline to provide the service or to resign as a director of the company. If the director is a member of a firm providing property development advice and the advice will be given to the company by a part of the firm not connected with the director and the director maintains they will not disclose any of the board's discussions relating to the service to those in the other part of the firm, it may be argued there is not a conflict. The argument that this type of 'Chinese wall' (i.e., a separation in an organisation designed to block the exchange of information between departments) is an effective protection is not accepted by everyone or in all circumstances and the director should consider resigning from the board.

In cases of conflict of interest, the practices to protect the director and the company are the same. The practices for dealing with a conflict or a potential conflict of interest are as follows:

- All potential conflicts of interest should be mentioned when a prospective director meets the Nomination Committee or Chair.
- When it is agreed that a candidate will become a director and join the board, the candidate should prepare a list of all their potential conflicts, noting the related organisation and reason for the potential conflict.
- The list should be tabled at the directors' first board meeting and noted in the minutes.
- The detail will be registered in the airline's 'Register of Interests', which should record each director's, and ideally each manger's, potential conflicts of interest.
- If a director identifies additional conflicts later the same declaration process should be followed.

- Regularly, probably once a year, the company's Conflict of Interest register should be reviewed with each individual director and any changes noted in the register and mentioned at the next board meeting.
- If a matter arises at a board meeting and a potential conflict becomes a real conflict, the director should remind the board of his or her conflict of interest, withdraw from the board meeting whenever the matter is discussed and not discuss the matter with any other directors, nor vote on the matter. The minutes of the meeting should note that the director withdrew from the discussion and meeting while the item was discussed.

Conflict of interest and the potential for a conflict is a matter for the whole organisation, not just the Board of Directors. Every decision-maker and certainly every person authorised to make commitments on behalf of the company must understand what a conflict of interest is and the procedure to be used when one arises or could arise. The procedure will be similar to the approach taken with directors and should be detailed in the company's manual of policies and practices, which will have been reviewed and approved by the board. The airline's internal audit function should frequently review the operation of the procedures on conflict of interest and report to the Board of Directors, probably via the Audit Committee.

Quorum

Before a board meeting can officially start its business and make decisions a 'quorum' (i.e., a minimum number of directors) needs to be present. The number of directors required before there is a quorum will be noted in the company's Articles of Association. The quorum required can vary for different meetings. For example, for a board meeting it might be 'two directors are present' or 'a majority of directors are present', whereas for a board committee it might be 'two-thirds of the committee members are present'. It is also possible to include some additional minimum requirements in the definition of a quorum. For example, 'A quorum shall be at least three members of the committee, provided that at least one independent non-executive director is present.' There is flexibility in setting the quorum for a meeting; the key point is that the requirement for a quorum is designed to ensure that, before each board decision is made, the recommendation received has been reviewed, discussed and voted on by more than one of the directors. This is consistent with the assumption which supports the operation of a board (i.e., that a group tends to make better decisions than an individual).

The statement of the required quorum frequently includes the phrase '… are present'. This leads to the question of what 'present' means in this context. Modern legislation tends to allow people to participate in a meeting and be considered as present even if they are not physically in the room. A board or committee member can attend a meeting by telephone or video-conferencing if this is permitted by the company's Articles of Association, but the participation must be in real-time so email communication does not help. A potential downside of directors or committee members attending a meeting electronically is the loss of personal contact and eye contact, but has the advantage that a director can join a meeting and contribute even if they are not free to attend in person because of other commitments. If the company's Articles of Association do not expressly permit directors to attend electronically, the Articles can be amended to permit it.

The potential problem of not having a quorum at a board or committee meeting can be overcome by a director appointing an alternate director (mentioned in Chapter 2) to stand in for the director when they are unable to attend a board or committee meeting. The alternate director has the same powers and responsibilities as the principal director. The company's Articles should authorise directors to appoint an alternate, or if not, they can be amended.

The assessment of the board and of individual directors

Most organisations have a system for regularly reviewing the performance of all staff and receiving feedback from them. The objectives of the system are to establish whether staff clearly understand what is expected of them, how well they are achieving their goals and also for managers to know whether they are giving staff the guidance and support they need. Usually this is done for staff using the chain of command, with seniors reviewing the efficiency and performance of those reporting to them, and the subordinates commenting on those observations and discussing with the manager the way they are being managed.

The Board of Directors also work for the company and need to make a contribution to achieving its success both as an individual and as a member of the board team. It is sensible that each director and the Board of Directors as a whole should also have their performance reviewed regularly.

Although the Board of Directors is as much a part of the organisation as any operating section or department there is a difference in that the operation of the board has to be assessed from two angles; how well each individual director is contributing to the board's work and how the board functions as a team. A further problem is that the Board of Directors is at the top of the company's hierarchy and they do not have a 'boss'. Thus, the board have to assess themselves. The board is responsible to the company's shareholders, but their only power is to remove a director. Also the shareholders do not attend board meetings so do not see the interaction of directors and how they perform. The Board of Directors has an important role in the company, being responsible for setting the company's long-term goals and guiding the company along paths which will achieve those goals. It is important that the board as a whole and its individual directors are working well.

Review of individual directors

A particular factor to bear in mind when assessing the contribution of individual directors is that a director is required to be a member of the board team working with other directors and at the same time need to take an independent view and to express it; just as staff are individuals, directors are individuals as well.

The assessment of individual directors is a more complicated task partly because of the process and partly because the assessment involves two different types of director; executive directors who perform an executive function in the company as well as being a director, and other directors who act solely as the company's directors. If an executive director's performance is reviewed by (say) the CEO or Chair, it can be difficult to separate how the individual is working as a company executive and as a director, but an attempt should be made and it may well be that an expert should be asked to design a suitable system for doing this. Given that promotion to being an executive director is usually based on merit and performance, if there is a problem it is most likely that an executive director is good at their day job, but less effective when functioning as a director.

A potential difficulty with the process of reviewing how each non-executive director is working is that each will have been invited to join the board and therefore would assume that their personality and approach are what the company wants. This may mean that suggesting improvements and changes could require tact and care.

Although directors do not have individual job descriptions there are established questionnaires which list questions which will help assess the performance of individual directors. It is likely that a questionnaire and guidebook to help the Chair lead in the assessment of directors will be available from the local Institute of Directors and it will include questions such as:

- Does the director actively participate in developing the company's strategy and monitoring progress?.
- Does the director contribute to the discussion of the company's financial and operating results?.
- Is the director knowledgeable about the company's business plans?.

Many boards give a simple questionnaire to each director to complete in respect of each of the other directors. The results can be collated by some trusted officer, possibly the Company Secretary and a summary given to the Chair as a basis for a review with each director. A face-to-face meeting between the Chair and directors individually, in addition to being an opportunity for the Chair and the directors to review performance, gives each director an opportunity to raise any matters that worry them with the Chair. It is useful for each director to also rate themselves using the same questions, as this provides a chance for self-reflection for the director. The assessment of individual directors can be a very productive and useful process.

This leaves the problem of how to assess the performance of the Chair. No-one should be excluded from the assessment process and the Chair has the important role of creating the team spirit and personality of the board. The assessment of the Chair can be carried out by giving each of the directors a similar questionnaire as used for other directors, but with slightly different questions emphasising the Chair's leadership role. For example, instead of asking 'Does the director actively participate in developing the company's strategy and monitoring progress'? the question might be, 'Does the Chair actively encourage all directors to participate in developing the company's strategy and monitoring progress?' The information in the questionnaire should be collated by the same trusted person. The results of the survey can be discussed with the Chair and all members of the board in an open meeting or privately with the Chair by another senior person, perhaps the SID or the Chair of the Nomination Committee, because they chair the committee responsible for securing the right skills on the board.

Review of the Board of Directors

There are many ways, informal and formal, for gathering opinions on the working of the board, such as:

- Allocating time during a board meeting for all directors to openly express their view on how the board is working.
- Having a meeting outside the board room, in a more informal setting, to discuss the board's operation.
- The Chair meeting directors individually to gain their views.
- A formal survey where directors assess the interaction and decision-making of the directors with the results summarised by an independent party.
- Inviting an independent expert to attend a board meeting and afterwards giving feedback on the board as a whole and separately on each director.
- Probably the most popular option, to ask each director to complete a simple questionnaire followed by a face-to-face meeting with the Chair on the individual comments.

But the objective is always the same; to gain a frank opinion as to how well the board works as a team, how well they co-operate and what improvements might be considered. Whichever approach to assessment is used it needs to be sensitive to the fact that it is dealing with people, hence will probably require quite a careful process concentrating on an open discussion in the same fashion as general staff assessments.

The most frequently used methods of gaining a director's view have been mentioned earlier. Generally, the board will assess how well they are working together, by giving their individual views to the Chair of the board. Any survey or discussion will include specific questions such as:

- 'Does the board spend enough time discussing, monitoring and explaining strategy?'.
- 'Are the board minutes a sufficiently clear record of discussions and decisions?'.
- 'Is the flow of information between the board and senior management appropriate?'.

A discussion between each individual director and the Chair regarding that individual director's answers will provide a good view of how directors consider the board is functioning. This can then be discussed by the board as a whole to find ways to improve, if improvement is needed. This type of process does not require the views of individual directors be published so this will/can encourage frank answers. One of the problems of developing the board as a team is that it is important for each director to have the attitude and ability to ask incisive questions and sometimes take opposing views.

In some jurisdictions companies above a certain size and/or whose shares are quoted on a stock exchange are required to have the performance of the Board of Directors assessed regularly, (say) every three years, by an independent assessor. These assessments are often made by the local Institute of Directors or professional firms. Most jurisdictions with this requirement also specify that the company's annual report should confirm that an independent assessment of the board's operation has been made.

Director and board assessments are important processes because they are part of making the board into a team and improving its effectiveness. They also provide a way to eliminate minor friction on the board and to build trust.

Education for directors

A director is appointed to the board because of the skills and experience they have and can call on for the benefit of the company. This does not mean that each director has perfect knowledge, as there always seems to be something new to learn.

A director needs to keep up their knowledge of two areas: the company's operations and recommended good practices for directors, as these two areas apply to all directors, even those who have been appointed from within the company. Both streams of knowledge, company operations and directorship, are equally important, but in the early days of a newly appointed director the priorities will be different. For a director appointed from outside the company, initially the top priority will be to learn about the company, while for someone appointed from within, the priority will probably be to learn about directorship.

Usually either the Chair of the board or the Chair of the Nomination Committee will be responsible for co-ordinating each director's education scheme. Naturally each scheme will be personal to the director and tailored to the information and knowledge they need. In some jurisdictions directors of public companies are required by law to attend designated governance and directorship courses within a certain time of being appointed to a board and in addition to spend a minimum amount of time each year on 'Continuing Professional Development', often referred to as 'CPD', which for a director is a continuous plan for acquiring new knowledge and skills in the profession of directorship. The education programme is quite flexible and can consist of:

- Attending courses, seminars and conferences.
- Reading books or material prepared by the company's advisors on governance and being a director and/or on specialist subjects like finance or on business affairs.
- General reading on directorship trends and current thinking.

Each director's scheme should be agreed in advance, by the Chair.

There can be a problem about who pays for the CPD. In a large company it is usual for the company to have a budget for directors' CPD, but this could be a burden for a newly formed start-up. Airlines are complex businesses with complex operations and it is important that all directors, including those appointed from within, learn how the airline works. In a large or small airline, the current executives can brief directors on how the various areas operate, as well as current problems, solutions and possible developments. Although this approach still has a cost, it is covered by existing expenditure and is really just an allocation of time; also educating directors should be time well spent. CPD covering general business developments, governance and directorship is usually provided by third parties who charge market rates, although the local Institute of Directors may run some courses which are free or have a nominal cost. If the cost of CPD courses are a burden for the company it may be possible to achieve the same result by buying books on the subject and the Company Secretary running short discussions on the aspects of governance after the books have been read. This is really back to the idea discussed before that governance principles can be achieved in a variety of ways that suit a company's circumstances. Directors have essential and important responsibilities and they need the knowledge to do their job well and to stay up to date.

There are good grounds for arguing that a director, or potential director, should receive training in good governance practices before they actually need it. It can only be helpful to give senior managers governance training because it will help them understand the importance of many of the governance matters they need to deal with and to administer. Examples are the need to avoid conflicts of interest and if a conflict arises how to deal with it, and the value of the statement of business ethics (discussed in more detail in Chapter 4). In addition, managers will have a better understanding of the reasons why the board requires information and why it is involved in the decisions on some matters.

Board meeting

Before each board meeting a copy of the agenda and supporting papers, recommendations and reports, should be sent to each director. This is usually done by the Company Secretary. Increasingly the board meeting package is being distributed electronically, but some boards and some individual directors prefer to receive copies in printed form. If the agenda provides that there will be a presentation to directors at the board meeting, a copy of the materials can be circulated with the board meeting package. The package for directors should be in directors' hands in sufficient time for them to read and absorb the information, and some boards aim to have the information with directors at least a week before the meeting.

After reading the information in the package, an individual director may have some questions and the company should have a mechanism for responding to them. There are two types of question, one seeking clarification of some information (e.g., 'why have debtors decreased by more than 10%?'), and the other questioning the basis or principles of a report or proposal (e.g., 'Why does the trend of forecast revenue differ from the trend in figures issued by the tourist industry council?'). There is no reason why questions which simply seek to clarify information should not be raised by the director and answered before the meeting and a frequently used approach is for the director's question(s) to be directed to the Company Secretary, who either passes them on to the correct officer or puts the director in direct touch with them. Questions on principles should really be raised at the board meeting so all directors can discuss the points made, but if the query is on a very major matter, it is wise to warn the Chair in advance.

At the meeting there should be an open and frank discussion amongst directors on all issues with the Chair acting as a facilitator and periodically giving a summary of the discussion to date. Once a conclusion has been reached, the Chair should sum up and re-state the conclusion so that it can be properly minuted. If there is not complete agreement amongst directors on an item or a decision, the decision supported by the majority should be recorded and the dissenting views noted, so that in this way the decision as well as the full range of views is recorded in the minutes. There may be circumstances when a dissenting director's contrary opinion is so strongly held and the matter is so important to the company that the director feels that simply minuting their view is not sufficient. In these circumstances the only practical course is for the director to resign from the board. This sort of circumstances is rare, but can arise, for example if there is a recommendation to change the external auditors for, what the director considers, to be the wrong reasons.

If a director has a conflict of interest on any matter to be discussed, they should withdraw from the meeting when the discussion starts and can re-join the meeting when the discussion finishes, with the timing of the director leaving and returning to the meeting being recorded in the minutes.

After the board meeting directors may take the board papers used away with them, but some companies require that all the papers are left behind with the Company Secretary to improve security. The company needs to have a secure system for recording and storing confidential information. Each director receives a considerable volume of commercially confidential information. The information is partly in written form or perhaps also partly in electronic form, but also in the director's memory. Each director is responsible for the security of the written information and most remain aware of the need for confidentiality when speaking to others; even conversations between directors can be overheard. The information issued to directors by the company remains the property of the company and when a director leaves the board, the information should either be returned to the company or destroyed. It is usual to discuss the disposal of confidential information with the Company Secretary or the Chair. It is not unusual for each director, before being appointed to the board to sign a 'non-disclosure agreement' which is an agreement not to disclose any of the confidential information on the company they have received while a director; this will apply to executive as well as non-executive directors.

Regardless of the differences of opinion expressed in the board meeting, once a decision has been made, all directors should support the decisions when dealing with third parties, which is often referred to as 'cabinet responsibility'. This may not always be an easy thing to do.

A director holds a position of substantial responsibility and it is natural that society requires directors to deliberate and decide on matters with care and caution and this desire is expressed in the various laws, regulations and recommendations which surround the director's position. The director needs to be clear on what the laws, regulations and best practice recommendations are and to stay in touch with changes proposed and made.

4 Strategy

What is involved?

For some people 'strategy' is one of those words where the reader's eyes glaze over and their concentration wonders off onto another subject. Yet the idea of strategy is a fundamental reason for having a business, it is the answer to the question 'What is the business trying to achieve?' Formulating a clearly understood strategy and monitoring progress towards achieving it is one of the absolutely essential functions of a Board of Directors, hence it is at the centre of good governance. There is more than one definition for 'strategy'. Creating and monitoring strategy is a difficult process to pin down and there is the danger that in an attempt to make the term understandable to everybody, words are used which make it harder for the average person to understand. In this book, as in its sister publication, *Airline Management Finance: The Essentials*, the definition of strategy is 'a high-level statement which describes and defines the long-term objectives of the airline which has been approved by the airline's Board of Directors; it is the airline's view of its future'.

Before exploring strategy further, it is imperative to make it clear that although strategy is a high-level statement, it is important that the essentials of the strategy can and are communicated to all levels in an organisation so that they are understood by everyone. It is the organisation as a whole which will achieve the agreed strategy, and the long-term objective, not just the Board of Directors, so it is important that everyone in the organisation understands what the airline is trying to achieve so they can work towards achieving it. This communication exercise can present the board with significant problems: problems which are particular to an airline, which will be discussed later. The key point to make at this stage is that a company's strategy needs to be understood by everyone in the organisation or it is unlikely to be achieved.

Definitions

Just as there are different management definitions for the word 'strategy', there are different words for the same definition. The description and definition of an airline's, or any other company's long-term objectives may be called its 'vision' or 'target' or 'goal'. The terminology or jargon, is not important; what is important is that there is a clearly stated set of objectives which the airline is seeking to achieve. At the formation of the airline the strategy should be agreed between the airline's founders and it probably is, but it is likely that the idea and the objectives are not written down, but are shared by the founders and communicated periodically and informally to the airline's managers and staff. Also, it is not likely to have been expanded to cover every function of the airline. Strategy is a statement of how to get to the desired long-term goal.

Developing and formalising a strategy

Perhaps the original idea for forming an airline might be the thought that: 'There are many unused evening and night take-off slots at regional airports. If the airport authorities will offer cheaper rates to use those slots, it should be possible to attract passengers to travel at unsociable hours.' If, after discussions with a number of regional airport authorities it seems there is a strong possibility that they will offer lower rates to induce airlines to use unpopularly timed take-off slots, the idea may be expanded into a full business plan. At this stage the 'strategy' or 'goal' or 'vision' is 'To provide passengers services from regional airports which have unused inexpensive take-off slots' and this idea should be written down as a reference point. It may be that as the basic idea develops, the statement will be amended or expanded, but the initial statement provides a starting point. From this point a business plan can be developed which describes how each part of the airline will operate to achieve that goal. Discussion of the overall strategy and questions seeking to clarify what this will mean in the airline's daily operations will produce strategies for other parts of the airline. For example:

- 'How will we attract people to fly at unpopular times?'. The answer may be 'low fares and to give every passenger access to a lounge and provide subsidised quality food before take-off'. If this answer is accepted there are elements to include in the Marketing Strategy. If the approach is to use the lounges of other airlines when they don't want to use them this will be included in the Operations Strategy.
- 'Taking up slots which are currently not used' probably means there will be a peak demand during the period of school holidays and the regular demand will be Mondays to Fridays for business men and women with little traffic at the weekend. If this is a valid assumption the fluctuations in a regular demand (only on working week-days) and seasonal demand (school holidays) will influence many of the other strategies. The Financial Strategy will need to cover how to manage fluctuations in cash flow, the investment of surplus funds and the possible need for a short-term borrowing facility. The Operations Strategy will deal with how to ensure the availability of aircraft, permanently and temporarily, and how to organise the use of other airlines' lounges as well as check-in facilities and the training of cockpit

and cabin crew. The Engineering Strategy will need to detail how maintenance will be scheduled and consequently how the level of spares and stocks will be managed. The Strategy for People should include the recruitment and retention of permanent staff, how to access temporary staff, their conditions of service and training. The Information Technology Strategy will deal with the fluctuating demand in using the support systems during the various seasons as well as testing and introducing new systems during slack periods. One simple statement can have effects through many of the supporting strategies.

The iterative interrogation process will result not just in strategies for each part of the airline, but strategies which further clarify how the overall strategy will be achieved, and are also a basis for designing measures and milestones to gauge progress and to determine efficiency. These measures and milestones will be the all-important Key Performance Indicators which are discussed a little later. These plans may also be called strategies (e.g., Marketing Strategy), or may be called tactics, but essentially, they describe the way each part of the business will support the overall strategy (or whatever other term is used). Tactics are the carefully considered and agreed ways to achieve the airline's long-term strategies.

Although in theory an overall strategy and the supporting strategies should be developed when the airline is first founded, this does not happen in all cases. A common approach is for the airline to start operations on a small scale, learn the lessons from the operations, then if the business is successful, build on the experience. Perhaps at a later stage there will a suggestion that the overall strategy and detailed supporting strategies which have been developed on the run but not written down, should be reduced to writing and agreed by the Board of Directors. If the business is not successful then the founders will either go back to the drawing board or examine other types of business. The experience of the first operation will give the founders more confidence when describing the overall and supporting strategies.

Frequently the reason why an airline does not consider describing its strategies until it has been operating for a number of years is because most of the managers and executive directors will have been with the airline since its foundation and had frequent contact with the founders, hence they understand from practice what the airline is trying to achieve. Once a significant number of new staff join the airline, the new staff need to understand the airline's focus and a written description of the overall and supporting strategies will tell them this.

Discussing the various circumstances in which strategies are developed highlights the variation in timing that there might be, as some airlines may establish strategies when the airline is formed, while others go through the process much later. It is worthwhile considering the possible effects of waiting too long to formalise the strategies: are there dangers for the airline in not having a defined reference point? Could the airline start investing in too many of the different parts of aviation? How will decisions on new major investments be made? Whichever course is taken, the responsibility for approving the overall and supporting strategies rests with the airline's Board of Directors. It may be that the wording of the strategies is developed by others (e.g., consultants or the airline's planning group), but the responsibility for interrogating, debating and approving the final document rests with the board.

An important part of the set of strategies is agreeing the measures and milestones to be applied to them. It is likely that some key words in the strategy such as 'inexpensive' or 'efficient' express a good intention, but are difficult to translate into management action day-to-day. To bridge the gap between the intention and the execution, the working guide to each section of the strategy should have measures or ratios against which the airline's operations can be compared (e.g., 'inexpensive take-off slots' might be defined as any slots which are 30% below the standard tariff, and 'efficiency' could be described as having a cost per seat mile which is less than 80% of the industry average). This approach is not designed to impose impossible targets, but to define the strategy in a way that can be easily understood by the airline's operating management and used as a comparison with actual performance. Similarly, the milestones give a timescale for the strategies. For example, in the strategy for the customer there will be some measure for customer satisfaction together with target dates for achieving improvements; it would undoubtedly be debated by the board as to whether the airline should aim for 100% customer satisfaction at some stage.

Resources and planning

Strategy is closely linked with planning and resources. The strategy package is the basis for more detailed plans and these plans generate a list of resources needed. Just as the Board of Directors is responsible for ensuring that a set of strategies is developed and communicated, it is also responsible for ensuring that the organisation has the resources it needs to achieve the goals, when they are needed. It is tempting to think of resources solely as being money, but there are other areas which have to be considered and they fall under four main headings:

- Financial: this usually comes first on the list because securing all the other resources almost always requires money in some form. The financial resources are made up of two parts: finance currently available, which includes funds in the airline's bank accounts and invested in readily saleable investments (e.g., bonds); and the second resource is the ability to raise funds in the future through issuing more shares and/or increasing borrowing and/or re-investing profits earned by the airline.
- Physical assets: these include aircraft whether owned or leased, stocks of spare parts, premises and all the other equipment needed to operate a business; the responsibility includes not only to ensure the current needs are met but also the assets needed in the future will be secured in time.
- People and skills: a difficult but essential area, which always needs the close attention of the Board of Directors and their attention is needed whether

the airline is small or large, start-up or mature. Almost inevitably an airline's people will be involved in delivering the airline's product to passengers and shippers, so in many cases they are seen as being 'the airline'.
- Systems or technology: which is extremely important to any business operation and has the potential to improve the quality and efficiency of an airline's operation, but it can also harm an airline's reputation if not used well and effectively.

These four basic resources are sometime further analysed into sub-sections but these four are the essentials. It will be noticed that each of the headings requires the board to look into the future because almost none of the required resources can be acquired quickly, except in exceptional circumstances like the liquidation of another airline. The integration of the new resources also needs to be planned. For example, adding a new aircraft into an airline's fleet requires detailed planning: where the aircraft will be used, availability of sufficiently trained onboard crew, integration of the aircraft into the airline's maintenance programme, the need for additional stocks of spares and so the list goes on. But this type of forethought is also needed for all the other areas. If funds are to be raised by issuing new shares, a great deal of planning is needed. The development of new customer support or technical support systems needs to be thought through and prioritised. Also, the recruitment of skilled staff or training of existing staff needs to be considered in advance. The board always needs to have its eyes on the future.

Circumstances change cases

Aviation operates in a dynamic business scene, competition changes as airlines enter or leave a market, customer expectations change in response to competition, not just from airlines, but also other means of transport, and a change in government policy in one country may increase or decrease market demand. There are really too many variables to list. The Board of Directors must be kept abreast of actual and potential changes in circumstances so they can evaluate whether any change or changes are needed to the airline's strategies. Examples of significant changes requiring serious consideration and possible amendments to strategies include: significant changes in fuel prices which may seem to appear permanent, or the importance of dealing with climate change. If a strategy is to be amended it should go through the same processes as agreeing the original strategies, ultimately being approved by the board and then communicated to the airline's workforce.

Key Performance Indicators

The measures which are used in the definition of the strategies should also be used to measure progress towards becoming a successful airline; the ultimate goal. They are often referred to as 'Success Measures' but more frequently as

'Key Performance Indicators' ('KPIs') and they are the principle set of measures a company uses to assess its progress towards meeting its strategic goals. If the airline's CEO had a monthly email of figures to show what is happening in the business whether in the office or while travelling, it would be the KPIs. They are the key pieces of information the board and senior managers need to see regularly, so they understand what is going on in the essence of the airline, or in any other business for that matter; KPIs are not used only in aviation. KPIs are the ratios or gauges which focus on what is happening in the essential parts of the business. The word 'Key' is important, as only essential and influential matters should be included in a set of KPIs. The KPIs will include both financial and operating measures together with some combining both financial and operating information and these can be classified as:

- Purely financial measures (e.g., return on equity).
- Purely operating measures (e.g., percentage of on-time departures, customer satisfaction).
- Combined financial and operating measures (e.g., average revenue per passenger, cost per seat mile).

To be effective, KPIs need to be:

- Relevant, can be either permanent, measuring an essential aspect of the business or temporary measuring a function which temporarily needs senior management's attention.
- Focused solely on important existential matters, which is not to say that ratios and measurers should not be used in other parts of the business, but the term KPI should be reserved for those which are truly 'key'.
- Able to reveal both bad and good results objectively.
- Calculated on the same basis in each period; if it is necessary to change the way a KPI is calculated, the figures for past years should be re-calculated using the new basis so that comparisons with past periods remain valid and any trend can be seen.
- Acknowledged by the airline's board and senior management as effective and efficient measures of the essentials of the airline's operations; measures which are not accepted by the management as being useful will be prepared, circulated, but ignored, hence a waste of time.

KPIs are an important and useful tool for a business of any size and a set of agreed KPIs should already be in place to monitor the airline's operation before the very first flight of a new airline. The KPIs can and should be modified as practical experience is gained, if a strategy needs to be amended, and when there are significant and permanent changes in circumstances. The greatest temptation is to add to the number of KPIs used and this risks the board and senior management being distracted from keeping the essential aspects of the business in front of their eyes. Before adding a new KPI, the board and senior management must be convinced that it truly is a vital measure by interrogating its usefulness, for example, by asking 'What will this tell us that we don't already know?' If the new measure really is vital it should be added to the set of KPIs and, if possible, the figures for previous years calculated; if the measure is not essential, but is useful, it can be added to the suite of other helpful measures reported separately.

An important use of KPIs is to highlight significant trends and that is why it is so helpful to have KPIs calculated from the first day and if a new KPI is added, to calculate it for past years. Frequently problems develop in small increments over time and watching the trend of each KPI is important. Every long- and short-term plan should include the projected KPIs which will result if the plan is implemented. The board must monitor the possible or probable trend of the airline's business and results.

In some jurisdictions it is required that every company include in its annual report a business review, including its KPIs together with a comment on them. There is often an exemption for any KPIs which, if published, might prejudice the company's business or reveal or indicate a business secret. The stated objective of this type of requirement is usually to enable members of the company (shareholders) and others (stakeholders) to understand the business. When there is legislation, it usually makes the Board of Directors responsible for judging which KPIs are included in the company's annual report. This gives some indication of the importance attached to KPIs and how carefully they should be constructed.

To review or not to review

The Board of Directors remains responsible for managing the company's strategy and supervising progress towards achieving its success. Therefore, they are responsible for keeping the strategic statements under review; circumstances do change cases. Part of the process of ensuring the strategy remains relevant is for the directors and senior management, separately and jointly, to periodically review all strategy statements to ensure they work in the real world. The frequency of the review depends on how the circumstances surrounding aviation are changing and the initiative to make a review should be taken by the Board of Directors.

INEDs can be particularly helpful in the process of reviewing the company's strategy especially if they have experience in related businesses. This does not just apply in airlines or parts of the aviation business, but businesses which are different but have similar features and problems as airlines. Examples of this are as follows:

- Should one of the INEDs have a background in retail banking, it may be they will have observation on ways for dealing with queues or online transactions or on using machines for routine customer transaction (e.g., Automatic Teller Machines (ATMs)) and the sorts of developments which may be in the pipeline.

- An INED involved in the fast-food industry could also have advice on queue management as well as ways to encourage passengers to pre-select their meal or refreshments, as well as the way standard elements of a meal can be combined to increase the variety of choice.

This type of approach can open minds to other ways of doing things.

Having a view of potential changes in the aviation industry is not the monopoly of the Board of Directors, as some staff within the organisation will have ideas on ways the airline could change. This is not a suggestion box for detailed ideas, for example, on ways to save paper in the office, but views on how the industry is likely to develop and whether the airline should consider changing its strategy. The move to automate the passenger check-in process could have developed from an airline's desire to keep its fares low (mechanising a manual process) or from a change in strategy to give passengers a range of options on how to check-in (i.e., online, at a machine or at a check-in desk).

Some companies may wish to have their strategy reviewed and audited by independent professionals, especially professionals with experience in aviation. This is certainly an option and does have the extra ingredient of having an independent view on the strategies.

Care is needed with the external and internal strategy sessions to ensure that they focus only on the key parts of the business and that any recommended changes to strategy are reflected in all related strategies. It is not helpful to recommend that the Customer Strategy be amended to giving the passenger the widest range of ways in which to handle their travel (e.g., three different ways to check-in), if the airline's financial resources are fully committed in the purchase of new aircraft to meet the objectives of the airline's Environment Strategy. Priorities need to be set by the board.

When it is agreed that a change is needed to the airline's strategy, the changes must be detailed in writing, approved by the Board of Directors and noted in the minutes of the board meeting, then communicated to the airline's staff.

Communication with…

There is not any mystery about what goes on at meetings of the board. A group of people with a variety of skills and experiences, meet to make decisions on what the long-term goals of the company should be, how they should be achieved, what progress has been made and what still needs to be achieved in the short- and long-term.

The decisions made affect the company, its shareholders, its staff, its suppliers and the community the company works in, collectively the company's 'stakeholders'. Historically the attitude was that the only group interested in the company, its operation and future, other than the board, were the shareholders. At that time the main method of communication was the company's annual report and a once-a-year meeting between the Board of Directors and the shareholders. These requirements were included in the list of directors' legal responsibilities.

This attitude changed when staff became more organised and it was fully realised that as the shareholders have invested their money in the company, the staff have invested their time, skills and, to an extent, their futures, in the company. The responsibility to communicate with staff became a board responsibility really through recommended best practice guidance notes issued by professional institutes, frequently national Institutes of Directors.

The current attitude is that each company is being seen as a part of the national economy and society with a responsibility to consider the effects of its decisions on the economy and society. An illustration of the effects of these developments is that (say) ten years ago very few, if any, airlines included in their annual report any information on their:

- 'GHG emissions' (i.e., Greenhouse gas emissions).
- Use of utilities (e.g., electricity and water).
- Donations and assistance to charitable organisations.
- Support to those of their staff who wished to do charitable work.

It is probable that the reporting for all companies, especially high-profile companies with a significant interaction with society like airlines, will, voluntarily or by legislation, continue to widen the scope of their reporting and communication. These developments are likely to generate changes in the detail of each company's strategy and the definition of a business' 'success' could well be changed and expanded.

Active communication with all those interested in the airline is now an important part of a board's responsibilities and it is an area which is likely to expand. To communicate effectively an airline's Board of Directors should develop a policy for communications, and fortunately there are now many more ways to help directors communicate with shareholders, staff, suppliers and society than there were 20 years ago.

'Communication' has more than one definition but an important point is that communication is not complete until all the parties involved in a discussion have the same understanding of the subject and the message. This does not mean that everyone agrees with the decision or strategy or conclusion, but everyone has the same understanding. Making an announcement is not communication; it is the first stage, and only a start to the process of communication. For a message to be understood and fully communicated, it will probably be necessary for there to be more than one series of explanations, exchanges of views or question and answer sessions.

Where there is a legal requirement to make an announcement or to issue a report there is always a permitted exception. A company is not required to release or publish any information about its secrets (e.g., a process or plans) if they are commercially sensitive and revealing the information could damage the company or its operation. For a company whose shares are quoted on a stock exchange, there is usually a requirement to issue a public announcement if the company is contemplating a major capital transaction or if its results are expected to change significantly. This is so that all shareholders and those dealing with the company have this significant information. Although the requirement is usually applied only to publicly quoted companies, it should be borne in mind by other companies as well because it may be something they will want to consider doing voluntarily.

...other directors

Communication between directors should be a comparatively easy process to establish and control. The board is a smallish group which meets face-to-face periodically and the Chair is responsible for ensuring that everyone has their say during board meetings and that all decisions are made after appropriate discussion and debate. If directors – executive, non-executive and INED – are expected to communicate with each other outside board meetings, it will be advisable to establish some rules to avoid confusion and possible irritation. Usually the rules can be as simple as requiring that all communications between directors are routed through the CEO or board Chair, although there are those who argue that this puts too much influence in the hands of one person, and that all communication between directors on board matters should be copied to all directors. Some form of compromise would need to be found.

...shareholders

If an airline starts with a small fleet, it is likely that the director's communication with shareholders is frequent and will cover any aspects of the operation the shareholders wish to review. As the airline grows it may be difficult to maintain this level of communication and the difficulties can increase if the airline issues its shares to the general public because not only will the number of shareholders increase substantially but the span of shareholders' interests will expand as well. For example, some shareholders will be primarily interested in the dividends the airline pays, some the potential for an increase in the airline's share price, and there is also likely to be a large number of individual investors as well as professional investment companies, possibly in addition to the founders. This can produce a mishmash of interests and groupings.

The national company law is likely to specify how frequently the directors should meet the shareholders and what type of formal reporting a company must send to its shareholders. This may be supplemented by the requirements of a stock exchange if the company's shares are listed and in addition there are likely to be recommendations by local professional institutes like an Institute of Directors. This legal framework forms a base-line, but the company may do more if it wishes, and there is encouragement for companies to do more than simply report to the shareholders the bare minimum of the financial result and financial position. The trend is to move away from producing a mainly financial report with comments on the business to a comprehensive report which describes all aspects for the company's operation, mentioning the current condition of the business; not just its finances, but also the commercial and operating situation together with a discussion of the board's assessment of the company's opportunities and the risks it faces.

The expansion of company reporting is likely to continue to develop steadily because purely financial reporting does not give a comprehensive picture of the company's operation and its effects on the economy. The scope of company reports will continue to expand and be refined, and a challenge will be to ensure they are easy to read and understand. Some companies have split their annual report into separate short sections focusing on individual parts of the business (e.g., 'Environment', 'Strategy', 'Finance'). This approach has the advantage that a reader can quickly find the information they want without having to read the whole report. Another approach is for the company to include on the company's website the detail of the information which does not change frequently, such as the way the Board of Directors operates, and simply include a reference to it in the annual reports. These changes make it clear that statutory reporting is an exercise in communication.

Many companies have set up a dedicated channel on their website which shareholders can use to ask questions and seek clarifications of the published information. The message usually goes to the company's Communications Department or Investor Relations Department. If this option is used, it is important that each message from shareholders is treated individually and that the temptation to use standardised replies be avoided. The degree that directors need to be involved in formulating responses to shareholder questions and comments will depend on the board's communication policy and, as a practical issue, the volume of messages received.

In association with the shareholders' access to company information via its website, a company can send information to shareholders on current developments and plans, leaving the way open for shareholders to comment or query if they wish. It is easiest to do this electronically, but care is needed because not all shareholders will want the information in electronic form and it may be necessary to produce a written circular. In addition, an email address list for all shareholders will need to be kept up to date as will the list of postal addresses.

...staff

Communication with an airline's staff is a different problem. The staff are involved in the routine operations of the company, hence usually have a more focused view. Effectively communicating with airline staff on the airline's strategy, plans and what has been achieved is not always easy, but the board is responsible for establishing methods of passing information to staff.

There are two main problems the directors need to consider when thinking about staff communication. The first problem is that many of the staff do not work in the office (e.g., cockpit and cabin crew, check-in staff). In addition, there will probably be members of the airline's staff working in locations away from its head office, in other cities and in other countries. In addition to all these absentees, some of the airline's functions, including parts of the airline's 'product' may be sub-contracted to other companies (e.g., check-in and ground handling). Despite the difficulties of communication these staff are essentially part of the airline's product and they need to know what is expected of them and what the current situation is, so they can make decisions which are consistent with the airline's strategy. A sub-set of this problem is that there may be a high turnover of staff in some areas of the business (e.g., airport passenger handling), therefore there is a continuing need to bring new staff up to speed.

The directors also need to consider why it is important to brief all staff on strategy, results and plans. A pertinent question is 'How will this information help technical staff, (say) an engineer, in an overseas airport do a better job?' The answer is that it probably won't, because they are already qualified, competent and motivated. The benefits of communicating with all staff including those who are absent are:

- To reinforce the fact that the staff are part of the 'airline's team'.
- To explain the reasons and background for decisions, even if the staff do not fully understand all the reasons for a decision, explaining the background to it and the process for reaching the decision which will give staff confidence in the management.
- So that information can correct rumours.

There is not a simple answer as to how best to communicate with all staff, even with all the electronic options available. It is easy to make announcements, but more difficult to have an exchange of views or to hold a question and answer session and to introduce the personal connection which can be so effective. The range of options available to directors include:

- Issuing a circular or briefing booklet, electronically or physically.
- Using an internet chat room.
- Circulating CDs with a presentation and FAQ (Frequently Asked Questions) section.
- Organising meetings with groups of staff. This can be difficult for staff who work on a roster and will ultimately mean that attending the meeting becomes part of the staff's duty time. This approach can work well with smaller airlines where it is possible to get the staff into one room.
- There is variation of the meetings with groups of staff and that is a form of 'Town Hall meeting', which is an open meeting to which all available staff can attend. These can be arranged so that over a six-month period all staff are given the chance to attend. The speaker at the meeting would generally be a director whose main purpose will be to explain strategy, results and priorities and, equally importantly, to encourage and answer questions.

- The board or a director having a detailed briefing with senior managers including a period to ask questions, after which the senior managers give a similar briefing to their department's managers and staff within a day or so.

None of these options are perfect and directors should speak to groups of staff when they periodically visit departments in the airline's head office or in other domestic or overseas locations. It is important to create an atmosphere of open communication on the airline's strategy and its progress towards achieving it.

The question of who should speak at the briefing sessions is often discussed and there is a wide variety of opinions:

- One view is that the responsibility for speaking to staff should rest with the Chair of the board or the CEO because they have the widest view of the airline as a whole organisation and can discuss any aspects of the business.
- A competing view is that it is better for the Chair or CEO to give an outline of the position and for executive directors to speak on the subjects which fall within their area of responsibility (e.g., the CFO on the financial position), on the assumption that they understand the detail better than anyone else.
- Yet another view is that the decision as to who speaks should not be driven by the subject, but by the audience. Hence the director responsible for flight operations should speak to cockpit crew and flight operations staff, on the basis that the information will then be explained in a way and with words that the audience will easily understand.

All of these approaches have their advantages and disadvantages and the board will need to decide which approach to take or how to mix the approaches. The key point is that the Board of Directors needs to be responsible for communicating with the various audiences and for deciding which is the best approach.

Achieving the required result with the staff of sub-contractors or agents is much more difficult, particularly where the sub-contractor is another airline. Probably the best that can be done is for a director or senior manager to meet the sub-contractor's staff who work on the airline's services and discuss in general terms the current position and plans of their airline.

The second problem for the board to consider is what information should and can be given to staff. Not all staff can or want to learn how to read a Profit and Loss Account, Balance Sheet and Cash Flow Statement and in any event the airline's finances are only part of its picture. This is where KPIs can be very useful, particularly those which are a combination of money and operations (e.g., cost per seat mile), and those which relate to the product (e.g., customer satisfaction). By looking through the airline's strategies, all of which will have KPIs, and selecting perhaps a maximum of half a dozen of them, especially those which are absolutely key and easily understood, it should be possible to discuss with staff the key area of strategy, results, the current position and plans.

If a company's shares are quoted on a stock exchange it is likely that the exchange's rules will require that any 'price sensitive' information (i.e., information which is likely to change the price of the company's share price), should be available to the general public rather than a small group like the company's staff. This can limit what matters the directors can discuss with staff, but having the discussion with staff using the company's KPIs may be a way to communicate the essential message to staff without breeching the stock exchange's rules.

Another area which the Board of Directors needs to consider is the extent to which some aspects of the financial reporting need to be explained to staff. For example, many staff may find it difficult to fully comprehend the effects of changes in foreign currencies on an airline's results and financial position.

....other stakeholders

This is not one homogenous group and the composition will change depending on the exact interest the group has in what the airline is doing. There have been examples of a group of stakeholders taking an active interest in a particular project, perhaps building a new hangar, who become much less active after the decision has been made and implemented. A company's stakeholders are important because they are groups which have an interest in what the airline is doing and what it plans to do. In some cases, a group of stakeholders may seek to influence an airline's plans or seek to influence the priorities of the whole aviation industry. An example is assessing the effect of the airline industry on climate change.

The list of stakeholders can include customers, current and potential suppliers, current and potential lenders, trade unions, government departments and local government councils, and again, the list can be rather long. There are techniques for identifying the groups of stakeholders interested in a particular airline and establishing each group's particular interest. The work is usually done by the airline's Communications Department or by specialist consultants. The board's responsibility is to ensure that all stakeholders are identified and their actual and potential interest in the airline is identified and then to act on the outcome. Identifying groups of stakeholders is not a one-time project and the information needs to be refreshed periodically.

Deciding how to communicate with stakeholders can be difficult and, in most cases, companies try to meet stakeholders' information needs by expanding the scope of the annual report. The approach may apparently solve the problem, but may make it more difficult for some other people to read and understand the annual report. In turn this problem has been solved by some companies splitting their expanded annual report into smaller sections (e.g., Strategy, Finance, Environment, Operations), so it is easier for people to find the information they need.

There are some organisations whose Boards of Directors have decided to publish extracts from the minutes of their board meetings as a way of communicating with the public in general and other groups like staff. This approach currently seems to be used mainly by government corporations, utilities and some large public companies, but there is no reason why a commercial company should not also consider this as an option. When minutes are published, they are almost always edited to some extent by removing purely routine administrative matters, but more importantly omitting any item which is commercially sensitive.

In general, communication with stakeholders is through the airline's annual report, other periodic reports and the airline's communication channel on its website.

Business ethics

Business ethics, sometimes called 'corporate ethics', is closely related to a company's strategy and is at the core of corporate governance. Some companies have opted to discuss and define the company's ethics in a written form by issuing a Code of Conduct. This is not yet a universal practice, but is becoming more popular. Many stock exchanges have a standard Code of Conduct covering all the companies whose shares are traded on the exchange and a company may simply adopt these or agree to follow them without issuing its own code.

The law requires that a company shall do and shall not do certain things (e.g., it must accurately describe what a product will do and it must not make any false claims for its product). The law will also provide for penalties if the law is not followed and this is much the same as the way the law governs an individual's behaviour. However, the law provides only one guide as to what is acceptable behaviour, and society in general has additional standards of what is acceptable and what is not acceptable. If we consider the question, 'what is acceptable behaviour?', the law provides part of the information and is in written form, while society's customs provide the rest of the answer and is not generally written down. Society's ideas of what is acceptable behaviour can change over time and directors need to be aware of the changes and decide the extent to which their company's code should be amended. What is acceptable can be a moving target.

An effective Code of Conduct can improve the trust of staff, customers and suppliers in the company and give others the confidence to deal with the company.

The Board of Directors is responsible for developing, approving and reviewing the company's ethics. Naturally the board is also responsible for devising methods to monitor the company's operation to establish whether the code is being followed. Frequently this is a by-product of internal audit reports, but having a secure, confidential method by which staff and suppliers can comment to directors or senior management on potential breeches of the code is very important. This is frequently called a 'Whistle-blower Policy'. It can also be useful for the board to periodically survey staff and suppliers to try and establish whether the code is being followed. To be effective the code must apply to all levels in the company including directors, be seen to be supported by directors and senior managers, and clearly state the consequences of anyone breeching the company's code.

The areas covered by a Code of Conduct usually include:

- People: such areas as equal opportunities, bullying, fair wages and human rights.
- Finance: fraud, tax evasion, keeping accurate records, not using confidential information for an individual's personal gain ('insider dealing').
- Safety: ensuring staff work safely and that all products can safely be used.
- Business: how to deal with the difficult area of employees giving or receiving gifts and hospitality needs very clear and detailed guidelines which will apply to everybody in the company including staff, managers and directors, and it may well rely on a local law governing this area. The code will also cover conflicts of interest, having an open tendering process, not using company resources for personal use and the degree to which the code should be applied to suppliers and sub-contractors.
- Information: protection of staff and customer information, not to mis-use information or collect unnecessary information.

These examples illustrate the very wide range of matters which can be included in the code, none of them revolutionary and most being common sense, but stating them in the company's code reinforces the idea that people's behaviour when working for the company should be at the same high standard as in their private life. One of the arguments used against having a Code of Conduct is that it merely reminds everyone to behave as they usually behave and use their judgement as they do every day.

Where a Code of Conduct exists, a company should comment on it in its annual report noting that the code exists and whether the board considers the code is effective.

Once the Code of Conduct has been finalised for the company, the board should consider the extent to which the company can or should require its suppliers and associates to adopt a similar code. In recent years there has been a great deal of pressure on companies to ensure that their suppliers and all the companies they are associated with operate to the same high standards as the main company. This is not always as easy to achieve as it may seem. Some suppliers may see this request as an insult and/or an attempt to interfere in their business affairs and decision-making. There may be even greater difficulty where the suppliers or associates operate in a different country or jurisdiction or where there are different legal requirements or business practices. The discussion on such issues as fair wages, safety at work and paying or receiving gifts may be protracted and that will be before the matter of how the company will monitor compliance with the suppliers' Code of Conduct or agreement to follow certain practices. Patience and persuasion may be needed to encourage changes in the suppliers' or associates' operations. The company needs to remember that if a problem does develop in a supplier's operation it may well affect the company's reputation amongst its customers, other suppliers and staff.

Crafting, approving, communicating and monitoring a company's strategy, together with supporting strategies for each area of the organisation, is a core responsibility of the Board of Directors and an essential part of corporate governance. It is not an easy task because the future is unknowable. Creating a Code of Conduct which discusses the business ethics of the company is a very helpful support to the strategy.

5 Reporting

Information

In the earlier chapters there has been discussion of the legal and good governance requirements for the Board of Directors reporting to and communicating with shareholders, staff and other stakeholders and of the pressures to do this well and to a high standard. The directors are responsible for managing the future of the company even though the future is unknown and unknowable. Part of the process is to have agreed strategies which are based on good information and then monitor progress towards achieving the strategic goals. In most companies these days it is probable that the majority of the board will be non-executive directors, so there is a need for all the directors to receive regular and focused reports on what is going on in the company and in the industry and be briefed on trends in the general economy and society. Although each executive director will know what is going on in their area of responsibility, they may only have a general idea of what is happening in other areas of the company or in the industry in general. All of the directors need information about the airline's operations, finances, business risks and operating environment. It is worth repeating that the directors need information, that is useful data they can use as part of decision-making, not just volumes of raw data. Whatever reports are given to the board they must have useful actionable information.

Before discussing the information that should flow to the directors, it is useful to identify one area of the business which is particularly important for all directors and it deals with the future and is an area where information is vital, and that is 'solvency'.

Solvency

A working definition of 'solvency' is the ability to pay debts as they fall due. It is not necessary to meet all liabilities at the same time, unless that this is the way the company has structured its finances, but when they become payable in the normal course of business. The reason this area is of such importance to directors is because the inability for a company to pay its debts will frustrate its ability to meet its goals. In addition, the Board of Directors is legally responsible for ensuring that a company remains solvent. There can be legal consequences for directors if a company becomes 'insolvent' i.e., unable to pay its debts as they become due. If the Board of Directors or individual directors are judged to have been negligent and failed to monitor the company's solvency or to consider the company's solvency position when making decisions, the directors may be fined, barred from being a director or become personally liable for any additional debts incurred, depending on the law governing the company. If a company continues trading while being insolvent it is usually referred to as 'wrongful trading' or 'trading while insolvent'.

In some jurisdictions the directors are required at regular intervals to review the company's financial position and issue a certificate that the company is indeed solvent. Where a certificate is required it is normally included in the company's annual report to shareholders.

Airlines are capital-intensive businesses and they routinely enter into long-term financial obligations. Buying a new aircraft means buying an asset which can operate for 20 years or more and whose re-sale value will fluctuate depending on fuel prices, market conditions, customers' perceptions and technological changes. All of these risks must be assessed when deciding how to finance assets which have such a long life. The most obvious examples of a long-term commitment are when an airline decides to buy an aircraft using a loan or to sign a long-term lease for an aircraft. In both these cases, not only is a long-term obligation created, but the airline's liquid resources will be reduced because of the need to make advance payments and deposits. There are other substantial commitments which airlines routinely make, for example:

- Buying or renting offices and on-airport facilities like warehouses and hangars.
- Changing the cabin configuration on its aircraft including new galleys, seats and inflight entertainment.
- Advertising campaigns to re-position the airline or support or change its branding.

Buying assets like aircraft or offices or hangars means that the airline owns an asset which in times of hardship can be sold. Similarly, if the assets are leased the airline can probably negotiate an early termination of the lease. However, in the case of re-configuring a cabin or a major advertising campaign, there are unlikely to be any assets that can be re-sold to raise cash to repay debts. It can be seen how cautious the Board of Directors of an airline needs to be when making, what for many airlines, are routine but large, decisions.

The main protections that a Board of Directors can take to avoid insolvency include:

- Having a long-term plan for the business which includes forecasts of cash flow and its financial position (Balance Sheet). The Balance Sheet will

show the forecast financial position of the airline and ratios calculated on the figures in the Balance Sheet will give an indication of whether solvency may become a problem in the future. Forecasts are, at best, educated or informed guesses. Hence, they are the best that can be produced at the time, but nevertheless long-term forecasts and the resulting ratios should be treated with some caution and not as firm figures.
- Regularly producing short- and medium-term cash flow forecasts and Balance Sheets, will give an earlier warning of whether solvency might become a problem in the future and will also allow the airline to calculate solvency or 'capital adequacy ratios' (i.e., ratios which seek to measure whether the airline has adequate resources − see Table 5.1). It is also important that the actual cash flows and Balance Sheet figures are compared with those which were forecast so that any significant differences can be investigated, system errors revealed and, hopefully, the accuracy of future forecasts improved.
- Calculating and monitoring the adequacy ratios and any others which can indicate whether a problem with solvency might arise in the future. The detail of the adequacy ratios is shown later in this chapter.
- Monitoring the airline's current assets/current liabilities position, particularly whether amounts due to the airline by all debtors are collected within the agreed credit periods and, if not, that effective action is taken to collect the debts. Also, that creditors are paid within the airline's agreed credit period, as failure to do this may indicate a shortage of liquidity (i.e., short-term funds). Finally, that other items in the airline's working capital, such as stocks of spares, and amounts due to creditors and 'Unearned Transport Revenue' (i.e., the amount of money the airline has received from passengers in advance of their travel), are all being managed and monitored against the company's plan.
- Ensuring that the value assigned to each of the assets in the airline's Balance Sheet is reasonable. This involves regularly assessing the value of aircraft, properties and, most difficult, intangible assets such as computer software.
- Ensuring the airline has a reasonable level of cash reserves to cover temporary cash shortfalls and difficult trading conditions. This helps to prevent creditors losing trust in the airline and possibly shortening their own credit terms, hence putting a squeeze on the airline's cash resources.
- The last element is perhaps the most important and the most difficult, which is to ensure the airline is profitable every year.

These practices and systems should not only help the board to get an early warning of any possible solvency problem, but will also demonstrate that directors have treated the matter of solvency seriously and consistently.

The principal ratios used to monitor solvency are presented in Table 5.1.

Table 5.1 Adequacy ratios (solvency)

Debt Ratio	Total Liabilities	÷	Total Assets
Debt/Assets Ratio	Total Debt	÷	Total Assets
Debt/Equity Ratio	Total Debt	÷	Total Equity or Shareholders Funds
Equity Ratio	Total Equity	÷	Total Assets
Quick Ratio	Current Assets	÷	Current Liabilities
Quick Ratio alternative	Cash + Debtors	÷	Current Liabilities
Solvency Ratio	Net Cash Inflow	÷	Total Liabilities

There is debate about the definition of the figures used in each part of the ratio and this is mainly influenced by the airline's accounting policies and also the way the figures are presented and described in the airline's statutory accounts. Confusion will be avoided, or at least reduced, if all reports on solvency define how each ratio is calculated and what is included in each of the figures used. For the first time the ratios are calculated, it would be useful to have them reviewed by the airline's external auditors. Although directors are entitled to assume that the figures reported to them have been correctly calculated, it is prudent for directors to critically look over them to ensure they are consistently prepared and that they indicate what the directors think they indicate. Monitoring the consistency and accuracy of figures reported to the board can be one of the responsibilities of the airline's Audit Committee.

The figures used in the calculation of the ratios are the total figures regardless of where they appear in the Balance Sheet. For example, 'Total Debt' includes both the 'Long-Term portion of long-term debts' as well as the 'Current portion of Long-Term Debts', and these two figures usually appear in different parts of the airline's Balance Sheet, the latter is usually shown as a current liability. There are also differences in the names used for each ratio, but what is important is not the name, but the figures used and that the reader understands what each of the ratios indicates. All of the figures used in calculating the ratios will be found in the airline's statutory accounts. Most of them are in the Balance Sheet (sometimes called the Statement of Financial Position) with some in the Cash Flow Statement (sometimes called the 'Statement of Cash Flows'). It is very useful for directors to look at the trend of the ratios because the airline's position may change slowly over time and looking at the trend is a good way to identify any worrying change. Any negative trend needs to be investigated and decisions made on how it can be corrected.

Information for the board

One of the Board of Directors' duties is to monitor the progress and performance of the company. To discharge this duty, directors need to feel comfortable that they have all the information they need about the airline's current and planned operations. There are four main ways for individual directors and the board as a whole to achieve this:

- Receiving regular reports on current operations and finances, and this is often the greatest volume of information a director will receive.

- Receiving irregular reports on topics which are of special interest for a limited time, like the introduction of a new aircraft type or the operation of a new passenger terminal.
- Receiving information on the performance of the airline compared to those achieved by its competitors and the aviation industry as a whole.
- Making visits to operating departments such as Engineering, and also locations other than the head office (e.g., offices in other cities).

The operation of an airline tends to be a complex business and it is easy for the board to concentrate only on its own strategy, plans and results, forgetting that every airline operates in a dynamic environment with competitors, some of whom are not airlines. There are other forms of transport (e.g., express trains), which may compete against short-haul airlines, and ships are competitors for some cargo carriers. It is important for directors to be able to put the airline's plans and performance into the context of aviation, as well as general transportation in order to see how well the airline is actually operating.

Directors have the right to request whatever information they believe is necessary for them to meet their obligation to monitor performance and solvency. They do not have to be content just with whatever the airline's management decides it will produce and circulate.

Volumes of information

Running an airline tends to be a management- and information-intensive job and potentially directors can be overwhelmed by the information available on its operations. Useful information can be difficult to find amongst tabulations of data. KPIs and important ratios can be particularly helpful to a director, making it easier to understand the figures and reducing the volume of information. A simple example is shown in Table 5.2.

Table 5.2 Ratios help the reader

	USD		
	Passenger revenue	*Passenger numbers*	*Revenue per passenger*
Year 1	20,532,558	28,966	708.85
Year 2	28,774,895	45,308	635.10
Change	+40.14%	+56.42%	−10.40%

The raw figures for Year 2 look very attractive compared to Year 1, however the 'Revenue per passenger' statistic shows that the increase in passenger numbers, 56%, has resulted in lower revenue from each passenger, a drop in excess of 10%: this is not an unusual occurrence in the airline business. This may be an acceptable result for the board and in line with the airline's plan, but the change in the 'Revenue per passenger' figure jumps out rather more than the revenue or passenger figures and should prompt some questions on the management of revenue and whether costs per passenger have decreased by a similar amount or percentage. In practice all of the figures and measures will be compared to the airline's short-term plan and possibly the measures will be compared to the same measures in the long-term plan. This is a simplified example illustrating the usefulness of measures and ratios in highlighting key information and reducing the volume of information. After a period of using the 'per passenger' figures as part of monitoring operations, the directors may prefer to focus on them rather than the raw total information.

This potential problem of decision-makers receiving an excessive amount of data exists in many businesses partly due to the complexity of business, but also there is a tendency for boards to ask for new information without stopping the flow of the old. Directors need to remember that producing reports, routine or ad hoc, costs money and uses resources within the company. To be effective, as well as cost-effective, all the reports and information flowing to the board should be reviewed regularly to ensure that it is still useful for monitoring the company's performance against plans. This may be a duty that the board will want to allocate to its Audit Committee, or, as in some companies, to the Company Secretary or CFO or Planning Department.

There is a general rule of thumb that it should be possible for a director to read and understand all the information they need for a board meeting within two hours. To permit directors to study all of the reports and information they need for a board meeting, often referred to as the 'board package', directors should receive it all at least a week, some say two weeks, before the board meeting. The information package can be sent to directors in paper or electronic form and it is wise to ask directors for their preference before adopting one or the other, and it may also be necessary to issue the pack in both forms.

Directors' questions

It is not necessary for a director to wait until a board meeting to ask questions. If a director's question simply seeks clarification of a figure or a ratio in the board package, it should be possible for any query to be directed to a named individual, perhaps the Chair or CEO or Company Secretary or some other designated person, who will refer the question to the appropriate executive to answer. The answer can be sent to the director via the designated person or to the director directly, but if this latter course is taken, directors should be discouraged from initially directing questions directly to executives. It is good practice for the questions and answers raised by directors outside the board meeting to be circulated to all directors. If the director's query is about a matter of principle or other important issue it should be made at the board meeting, although even in this case, it may be prudent for the director to inform the Chair in advance so that an answer may be given at the meeting. Some boards operate on the basis of 'no surprises at the board meeting'.

Frequent routine reports

The board's annual work plan, discussed in Chapter 3, mentions the reports that directors can expect to receive regularly before each board meeting as well the reports, plans and surveys which are often presented only at certain times of the year.

The directors can expect routinely to receive a report on the airline's operating and financial performance for the period since the last report, compared to previous periods and the airline's plan. The format for each part of the report is likely to be unique to each airline and will probably not use the same format as the airline's statutory accounts, however they are most likely to be prepared using the same basis and accounting principles as the statutory accounts. The names given to various reports will also differ. The reports will include:

- A Profit and Loss Account covering the entire operation of the airline. This is likely to be supported by a report of the profits and losses arising from the operation of all the airline's routes, probably prepared on two different bases. The first is, 'fully costed', where all the airline's administration expenses are apportioned to every route on an agreed, consistent basis to produce a profit or a loss for each route. The other basis is 'route contribution' where only the revenue and expenses which can be clearly identified with each route are shown and the surplus, often referred to as a 'contribution' is shown as contributing to meeting the airline's administration expenses. Supporting each schedule will be the operating figures, numbers of passengers, on-time performance, major operating problems, etc. The figures in each of these route reports will be compared to the airline's short-term plan and the efficiency will be measured by some statistics (e.g., revenue per passenger, cost per passenger, profit or loss or contribution per passenger). There should also be a forecast of the route's result for the financial year or next 12 months.
- A schedule of administration and operating costs showing for each manager the revenue and expense headings they are responsible for, the amount in the period compared to the short-term plan and the latest forecast. Any significant variations from plans will be explained. The directors will need to define the criteria for deciding what is 'significant'. In some organisations this is 'more than 10% and/or more than a value of USD'. The significance test should be applied to both the figures for the period and the cumulate figures because timing may produce compensating differences. The information in this schedule will also probably be supplemented by statistics. The differences between the actual result and the agreed short-term plan, usually described as a 'variance' should be explained by the manager partly so directors can understand the reasons for the difference. The discussion of variances and the reasons why they have arisen are valuable if they reveal

74   *Reporting*

whether there is anything to learn in order to make better decisions in the future.
- The airline's current liquidity (cash and near-cash) position showing the information in each of the airline's operating currencies as well as in the airline's home currency. The figures should be compared with the latest forecast with comments on any significant differences. This report will also forecast cash movements and fund balances for the following periods, at least to the end of the financial year and probably for the next 12 months. There will also be mention of the airline's current and proposed financial management transactions; the minutes of the Finance Committee will give the background to these figures.
- The airline's current financial position; also compared to the agreed short-term plan and any revised forecast. The summary figures will include statistics which measure solvency and the actual credit periods for debtors and creditors. Also supporting the summary will be the detail of the airline's working capital (i.e., debtors, stocks, etc.) together with measures of how efficiently each area is being managed (e.g., the age of debts, debtors as a percentage of revenue, the value of engineering spares per aircraft and so on).
- The minutes of each of the board's committees, which may lead to discussions on finance, audit, business risk and safety, etc.

The CEO, CFO and any of the executive directors on the board will explain the key information contained in the operating and financial reports and answer other directors' questions. There is sometimes a tendency for executive directors not to ask questions on the operating and financial reports, probably to avoid any implied criticism of a colleague. If the Chair has been successful in building the board into a team, all directors should feel free to ask any questions they wish without fear of offending anyone. There should not be any fear of questions amongst equals.

Infrequent routine reports

There are reports which are given to the Board of Directors only once or twice a year depending on how the airline is organised. Although the basic premise is that only important matter should be examined, discussed and decided by the board, it has the authority and the duty to decide what is important to the company. Hence each airline will have its own list of required reports. The most frequently required reports are:

- A proposed short-time plan, which may be called a 'budget', and which will include operating and financial information. It is likely to include a summary of the planned financial figures in the same format as the airline's statutory reports as well as in the format used to report the figures for each period. The financial reports are essential, but the money figures are the result of operating the airline. Therefore, it is important that the directors examine the operating information, the revenue plans and promotions, aircraft utilisation, load factors, numbers of staff expected to retire and be recruited and other key material and statistics before delving into the planned financial results. The short-term plan should also include a detailed cash forecast, analysed into months or quarters as the directors wish, and this report should highlight any planned major financings or financial restructuring. The amount and performance of the airline's reserve funds also need to be reviewed. Also included will be the planned capital expenditure together with a note as to whether it will be separately financed, and if so, how. The working capital part of the airline's financial position also needs to be reviewed in detail by directors to ensure that the amount invested is being used efficiently. Reviewing the short-term plan is an important task for the board, but adequate time should be allowed to study the information and interrogate it. This might mean having a separate meeting or being briefed by the senior management on the detail before the report is circulated and a decision is needed.

- Reviewing the airline's annual statutory reports and, if the airline's shares are quoted on a stock exchange, any report to the stock exchange. In many jurisdictions a company is required to issue statutory reports more than once a year (i.e., half-yearly or quarterly) and some stock exchanges have the same requirements of companies whose shares they quote and trade. Much of the detailed work on this will have been done by the Audit Committee, but they will, most likely, have focused mainly on the financial figures and dealing with any questions from the external auditors. The board as a whole also should review the figures. This leaves the board with the vital task of agreeing the content, wording, presentation and style of the business review to be included with the statutory reports. Producing a business review is a legal requirement in most jurisdictions, but everywhere there is a demand from society in general that companies explain what is happening in their business and this will include ESG information discussed previously. Annual reports have existed for many years, some estimate 50 years or more, and in recent years the readers of the annual report have criticised many companies for producing reports with figures and comments which are not related to the company's strategy or KPIs, also for including masses of data which can mask the vital information and producing reports which are difficult to navigate through for the ordinary shareholder. All of these are serious comments. The purpose of a report is to inform the reader – not to irritate them. In addition, it has been shown that the value of the company is often not purely a reflection of the figures in its Balance Sheet, but is in its brand and business model. While drafts of the business review may be prepared by a Communications Department or consultant, the Board of Directors is responsible for the content, wording, presentation and style: it is their report to their shareholders, staff and stakeholders.

- Review of customer satisfaction surveys, marketing tactics and competitors' performance. It is likely that this review will actually be a presentation at the board meeting by a senior staff member involved in marketing and sales, supported by a written report. In general, it is difficult to monitor revenue generation because of the variety of factors involved and the effect of real and perceived competition. The board needs to have confidence in the marketing plans and approach and to understand what is planned. Revenue generation is vital for an airline.
- Review a report on staff matters, such as staff turnover, morale, satisfaction and understanding of the airline's strategy. The report may also deal with staff pay policy. Reports on staff can be a very difficult area, particularly for non-executive directors, because staff do not always say exactly what they mean, particularly in resignation interviews which are an important source of information for assessing morale and satisfaction. It can be difficult to interpret the conclusions to be drawn from this type of report. If the airline has a large workforce it may be advisable for the report to be reviewed in detail by staff in the airline's Staffing Department at the board meeting. Life can be simpler for a small airline because the communication between the board, senior management and staff is probably less formal and the directors will have a better feeling for staff attitudes.

Other reports in the year

There are quite a lot of matters which the board needs to consider, but some are not particularly time sensitive and can be done when the board has time available in its schedule. They need to be looked at and reviewed, but that can be done when the board can give them the time they deserve. These matters include:

- Comparing the airline's performance with the airline industry in general and in particular with competing airlines (and other forms of transport where they compete with the airline). Airlines are management intensive and it can be easy for the board to concentrate only on what their airline is doing or plans to do, ignoring what is happening in the wider airline and transport business. One of the values of having non-executive directors on the board is to remind the executive directors and senior management to look outside the airline to see how other companies are developing, not just other airlines, but others in the transportation and customer services and safety-sensitive industries. There have been many changes in the airline business in recent years: some may be considered 'paradigm shifts' (i.e., major changes in the practices), others just better ways to do the same thing. An airline's board needs to monitor what is going on to assess the implications for their airline; the airline's future is their responsibility. Examples of major changes are, the introduction of 'low cost carriers' for short-haul flights (when will there be a workable low cost model for long-haul?), the move from booking through travel agents to individuals booking directly with the airline (with passengers printing their own tickets and boarding passes), the introduction of Business Class between First and Economy Class and more recently, Premium Economy Class between Business Class and Economy Class. There are many more examples and the board needs to have its eyes on potential developments.

- Reviewing the airline's policies and performance on the environment, staff, investor relations and risk management. None of these issues is unimportant and the board needs to find the time to consider what the airline is doing now and may need to do in the future, but the timing can be flexible enough to fit into the board's working plan. All of these issues are current 'hot topics' (i.e., matters which interest shareholders and staff and society in general).
- Re-examining the airline's strategy and consider whether any fine-tuning or major change is needed. This can be considered an 'ad hoc' (i.e., as and when necessary), matter and certainly the airline's strategy should not be amended too frequently, but it needs to be kept under review, as changes in circumstances, which are sometimes paradigm shifts, may require further thought on the airline's long-term goals and the tactics of how to achieve the goal. This review will also deal with such issues as the traffic rights the airline is currently using and those which are available but not being used.

The list of reports so far discussed cannot be complete, as aviation is a dynamic industry and the importance of issues can change. Each airline operates within its own environment and in its own way. The 'standard' reports discussed so far should be taken as an indication of the matters and types of topics the Board of Directors needs to be involved in. There will be other reports and issues which are specific to an airline or to the markets an airline operates in. The airline may have significant investments in airline related businesses, which might be operated through a series of joint-ventures, as each airline is individual and must report on and monitor each aspect of its business in an appropriate way, and the directors need to decide what is the best way. This is yet another responsibility on the board's broad shoulders.

'Prudence' and its application

Having read the detail of the range of subjects an airline's board needs to monitor, consider and decide on, it may cause concern amongst directors that they are required to be super-human. Perhaps they are, but that is not the way the law and best practice, and by inference, society sees it. In most jurisdictions a director of any company is required to act prudently, that is with care and thought in the way a prudent person manages their own affairs, avoiding speculation, seeking long-term returns and with the need to preserve capital. This brings the super-human back to earth as a mere mortal. Other words for 'prudence' might be, 'practical', 'careful', 'cautious' or 'farsighted'. The prudent director will always act in good faith for the good of the company and not in self-interest, seeking to promote the success of the company. The director will exercise the same amount of care in dealing with the affairs of the company as a prudent person, experienced in business, would use in their own affairs. This approach acknowledges that a director has responsibilities which are akin to a fiduciary duty, that is, a duty to manage and care for the assets of another party in their best interests.

The idea of prudence or caution should also be applied to financing, whether borrowing directly or indirectly through leasing contracts, or seeking to manage risks through financial instruments (i.e., a derivative). Directors, all directors, not just those on the Finance Committee, need to understand the risks, costs and benefits of every type of financial transaction the company enters into, before the deal is signed or agreed. Education on derivatives for executives should start early in an individual's managerial career and certainly before an individual gets to board level. Financial instruments are useful ways to change and 'manage' risk, but they can be expensive if the risks are not understood. Decision-makers should be told how derivatives work, the up-sides and down-sides, and how the exposure to risk is changed – not eliminated just changed. There have been many examples of companies which have suffered serious losses because they entered into financial agreements, loans or derivatives, without identifying and evaluating the risks and costs. There is recent experience of how smaller companies have lost significant amounts of money because they agreed to derivative contracts on interest rates without understanding or evaluating the risks involved. Small companies and start-up airlines need to understand derivatives. There are courses on the subject and gaining this knowledge is useful Continuing Professional Development ('CPD') for any director, but this sort of education should really be started when the potential director is at the managerial level in the company.

The idea of prudence also rolls into accounting and reporting standards, although there are differences of opinion within the accounting profession. The traditional approach to accounting and financial reporting is to take a cautious approach to the recording and reporting all transactions. This approach includes:

- Not over estimating revenue.
- Not underestimating expenses.
- Not over stating the value of assets.
- Not anticipating profits (i.e., not bringing them into account before they become real profits).
- Not revaluing assets upwards, but writing down over-valued assets.
- Making provisions for possible failures or losses (e.g., having a 'Provision for Bad Debts' as an estimate of debts that may not be collected).
- Valuing stocks and spare parts at the lower of cost or re-sale value.

All of these actions are part of taking a cautious, prudent approach. There are, however, many who argue that the cumulative result of being cautious and prudent is that a company's assets may be significantly under-valued and its profits under-stated. This counter-argument also argues that it is better to use estimates which are as realistic as possible for all values, revenue, expenses and capital items. To some extent current accounting standards incorporate the latter approach.

An airline's Audit Committee should debate and agree upon an approach which is sensible for the company to follow and use that approach consistently.

Accounting and reporting standards

For many years there has been a move to establish a uniform set of standards governing the accounting and reporting of a company's financial results and financial position. This project has been led by the International Accounting Standards Board and has resulted in a set of accounting and reporting standards which have been adopted in many countries of the world. Usually the standards appear as national accounting and reporting standards, but are actually the international standard perhaps amended for local conditions. The result is a uniform approach to recording and reporting financial figures.

Although there are international accounting and reporting standards that does not mean that every company or airline has the same accounting policies. A classic example is the calculation of depreciation, which is an accounting method of allocating the cost of a tangible asset over its useful life and reflects the decline in value from use. For an airline, depreciation can be a significant figure, perhaps running at 10% of operating costs. The accounting standard on depreciation requires that every airline should depreciate its assets as they are used. When calculating the amount to be charged in the company's accounts, all companies, including airlines will consider the same factors but individual companies may assign different values to those factors. The factors involved in the calculation of the depreciation charge include the cost of the asset, its estimated useful life to the company and its residual value (this may be its re-sale or scrap value). It should be noted that the useful life is the remaining useful life in the airline, not the period before the asset becomes unusable and has to be scrapped. It can be immediately seen that this calculation requires the airline's directors to make a number of estimates, such as 'how long will the airline continue to use the asset and will it be sold or scrapped, in which case what will its re-sale or scrap value be?' These are matters of opinion and will vary between airlines. For example, some airlines tend to keep aircraft for as long as it makes economic sense to operate them, while others have a policy of disposing of aircraft before they are (say) ten years old.

The local taxation laws may also indirectly influence the depreciation charge, but not always. Different jurisdictions can have different rules on what expenses can be deducted from income when calculating the profit on which tax is calculated. This means that an airline's profit for tax purposes may be different from the profit reported in its statutory Profit and Loss Account and a significant factor in the difference is likely to be the depreciation charge. The main tax approaches to depreciation by different jurisdictions are:

- To allow as a tax deduction whatever depreciation charge the company makes in its Profit and Loss Account.
- To only allow as a tax deduction, a depreciation charge which is calculated in accordance with a formula which is included in jurisdictions tax law regardless of the actual charge made in the company's Profit and Loss Account.
- To allow as a tax deduction, the lower of the actual depreciation charge in the company's Profit and Loss Account and the amount calculated using a formula in the tax law.

All of these differences in approach and opinions mean that the same aircraft type may be depreciated in different ways by different airlines and yet all will comply with the accounting standard. When deciding on how to value the factors involved, directors also have to bear in mind the arguments on prudence discussed earlier. These technical issues will be discussed by the Audit Committee and recommendations made to the board for approval. Nevertheless, all directors must understand the issues.

Once an airline has agreed its accounting policies, they should be used consistently for each accounting period and used in both the airline's statutory accounts and its internal management accounts. Using the same policy in each period makes it easier to make comparisons between periods. There may be occasions when an accounting policy needs to be changed (e.g., if the local accounting and reporting standard changes or because of a change in circumstances such as deciding to change the way it accounts for aircraft maintenance checks). Any proposed change must be reviewed by the board's Audit Committee and any recommendation to change approved by the Board of Directors. When a change has been agreed the effect on the financial figures should be noted in the airline's accounts and annual report and it may also be necessary to adjust previous years' figures to the new basis.

The airline will also have some accounting practices, which are convenient rules for how a transaction will be treated. For example, this may be above the value an item should be treated as an asset (i.e., an item which will benefit the airline for more than one year), rather than charged as an expense in the airline's current year Profit and Loss Account. As an example, the airline's board may decide any individual item of computer equipment costing more than a certain amount must be treated as an asset while any purchase below this amount is to be treated as an expense and written off as to the Profit and Loss Account in the month and year it is purchased.

Reporting finances and exposures

This chapter has discussed the director's responsibility for ensuring the airline remains solvent and separately the risks involved in dealing in derivatives. It is important for directors to be regularly informed about the airline's financial management, the current financial position and any planned financial transactions. If the airline has a Finance Committee they will be up to speed on the details and background to each decision and plan, but as in so many other matters, all directors need to understand the current position, what is forecast and what is planned and the possible implications. The easiest way for all directors to be informed is for the airline's treasury function to produce a report submitted to the Board of Directors via its Finance Committee.

Designing a comprehensive report for finances and financial exposures is not easy, as there tends to be a lot of detail and it can be difficult to isolate the important factors. Often the information can be most easily understood through graphs and charts. Another problem is that the report needs to be current. Financial markets can change quickly and the changes may alter the airline's plans for future transactions or its view of current financial positions. None of these problems is insurmountable, but directors need to be aware of them, so they get the information they need when they need it.

The contents of a finance report will vary airline by airline, but the financial position of a small airline can be just as complex and in need of the same amount of management as a larger operator. Essentially the report needs to include the current and forecast liquidity position, the current financial position, all forms of hedging, the current value of all derivative values and transactions, details of the future financing plans and the performance of investments. In more detail:

- A statement of all the currently held liquid funds expressed in the airline's home currency together with the reasons for any significant differences between the actual position and that previously forecast. There should also be a forecast of future inflows and outflows, usually for a year ahead perhaps split into quarters, also with reasons for any changes from previous forecasts.
- A similar statement in the original currency for foreign currencies together with a note on any restrictions (e.g., exchange control permission needed before the funds can be remitted). Also, of any self-imposed limits for example of the amount to be held to meet future expenditure or loan repayments.
- A list of all outstanding financial arrangements including derivatives, showing the nature of the agreement and their current market values. In addition, any planned financial arrangement should be noted.
- Detail of the investment of short-term funds and of funds held in reserve for emergencies, the amount and current return on the funds.

- A list of each financial institution the airline has dealings with, showing the volume of business and the value of unsettled financial arrangements. For a large airline it will also probably be useful to note any limit the Finance Committee may have set on the maximum amount of the value of outstanding contracts and other arrangements with each bank. In addition, the list should show the exposures the airline has from banks, so for example if the airline has a term deposit with a bank, that is an exposure the airline has to that bank.
- A list of transactions that the Treasury Department has authority to execute depending on market circumstances (e.g., to sell surpluses in a foreign currency if the exchange rate reaches a certain level).

There are other pieces of information the board will require. The important issue is that the report should be complete including all financial transactions and almost as a contradiction, the report has to be easily understood by directors.

Business risk

The world, including the business world, is inherently dangerous and risky and yet people survive and, for the most part, thrive. They do this by consciously, or unconsciously, continually identifying and assessing the risks they face, deciding what risks they can afford to ignore, what they can do to reduce the risk and what will happen if everything goes wrong. There are examples every day in everybody's life (e.g., crossing a road or deciding how to protect their house against fire). The Board of Directors is responsible for ensuring that the same process operates within the company.

Airlines have been safety conscious for many years partly because accidents can be disastrous for people and partly because airlines are required to be safety conscious by law and regulation. Every airline will have a safety organisation established to meet its licensing authority's requirements. Airlines also have business continuity plans to keep the business operating during difficulties like extreme weather or electrical failures or the hundreds of other circumstances that can make an airline's life difficult. In addition to flight safety and operating safety, there are 'business risks', which is a 'catch-all' term covering every other form of risk not within the definition of flight safety or operating safety and includes all potential occurrences or factors which have the potential to reduce the airline's chances of achieving its planned results or meeting its planned targets or its strategy and, perhaps to cause the airline to fail entirely. The Board of Directors is responsible for ensuring that the identification, evaluation and management of all risk is an active project for the whole airline because many risks, for example, those associated with the changes in legislation, never go away, but they may change in likelihood with a potential effect on an airline.

The director's approach is to ensure they have an overview of all risks facing the airline. Usually within the airline the flight safety and operating safety tasks tend to operate as separate functions, sometimes with different reporting lines within the organisation. To achieve a co-ordinated approach to risk, these three types of risk management operations – flight, operating and business – should be brought together, and co-ordinated through a Safety Committee or a Risk Committee. In smaller operations the responsibility for monitoring the three tasks may be allocated to the Audit Committee or even the Finance Committee. Although the reporting of risk may be funnelled into and through one committee, it is worthwhile appointing a senior manager, or a director to be responsible for risk, not necessarily interfering with the established lines of authority but being fully in the picture on flight and operating risk management as well as being responsible for business risks.

The process of managing risk is first to identify all the risk exposures the airline has, from internal systems and processes as well as from external factors. Once identified, then for each risk the process should be to:

- Define the risk and its possible consequences, if possible, assigning a value.
- Assess the likelihood of the risk occurring; it has to be remembered that it is possible for a 'once in ten years' occurrence to happen in successive years and for a 'once in ten years' occurrence to happen in the same year as a 'once in twenty years'.
- Assess the effect on the airline, financially, on its reputation and on its staff; when assessing the effects, it should be remembered how important an airline's reputation is and how difficult and costly it can be to repair a damaged reputation.
- Note possible future risks to be monitored.

It may be that this process, especially step two which is the assessment, can be helped by accessing the information held by external sources, for example, insurance companies and industry associations. The local Institute of Risk Management and/or Institute of Directors may also be able to help with information or guidance through the process. An airline with a small operation or a start-up airline should certainly consider initially trying to access sources of local knowledge and experience to possibly simplify the process for them.

With this information, a decision can be made on the importance of each risk and the priority for management to recommend the options for dealing with it. The recommended method of management can have effects everywhere and anywhere in the airline and may involve:

- Disposing of certain equipment.
- Buying new equipment.
- Changing the way some jobs are done.
- Extra training for staff.
- Changing the airline's insurances.
- Changing the basis for calculating the amount an airline needs to hold to meet emergencies.
- Considering the airline's relationships with major suppliers, for example, aircraft manufacturers or fuel suppliers to establish what alternatives there are.

The list is almost endless. The expertise in the airline's treasury area and/or the Finance Committee should be able to deal with the management of financial risks, but it should be borne in mind that if a piece of financial risk management goes wrong (e.g., the airline has liquidity in a currency which devalues substantially), it may be necessary to deal with the media and give details in the airline's annual report, and these considerations then add extra dimensions to the effect. It is for this sort of reason that all risk management needs to be co-ordinated through one committee.

Where the management of a risk involves a definite procedure, like dealing with a heavy landing which has elements involving ground operations, aircraft safety, passengers, finance and dealing with the media, it is worthwhile designing ways to check the efficiency of the procedures, testing them frequently and modifying the procedures based on the experience gained. It is also worth considering having tests with extreme conditions because there are no rules about what can happen. For example, a flight simulator can be used to simulate a landing in a typhoon or storm with one engine on fire and smoke in the cockpit, although the probability of this combination of circumstances may be very low.

Directors carry significant responsibilities and need to be informed of what is going on in the company as well as being given forecasts of the company's possible short-term and long-term future. The are many potential obstacles to a company achieving its strategy and the Board of Directors needs to have a picture of the current and possible future positions so it can give guidance and advice to the management. One of the challenges for management is to give the board the information it needs in a clear, concise and unbiased way. Information and reports are the life blood of the company.

6 Changing corporate governance with growth

The problems

Large companies and small companies tend to operate in different ways, but in most cases the large company started out operating on a small scale and became large because its business did well. It is also likely that part of the company's success and growth was due to being well run and well organised from its very early days. Corporate governance is needed and has an important place in a small company just as it does in a large company. Many argue that a smaller company, particularly a start-up, will gain more business benefits than a large company from ensuring it has an effective system of governance. Newly formed companies and operations need to establish a good reputation and have credibility with their customers, staff, financiers and suppliers. A practical and effective system of corporate governance will underpin an efficient operation.

The amount of time and effort spent on corporate governance has to be kept in proportion to the scale of the company's operations, as no-one establishes a company solely to have a good corporate governance structure. Corporate governance is a framework and structure to help the company define and monitor the progress towards achieving its objective. This is consistent with the approach that all directors need to understand the principles of good governance so they can apply them in a way which is appropriate to the organisation's size and current needs. One of the problems facing a growing company is to keep its system of governance under review and to modify it when change is needed. Using an example used earlier; at what point does a company change from having one of its INEDs, in consultation with the company's auditors, responsible for monitoring the quality of the company's accounting, to having an Audit Committee? The decision is one for the company's Board of Directors to make at the right time. To establish what is the 'right time', the position needs to be reviewed regularly and objectively.

There are many ways for an airline to start its business. They range from starting on a small scale with an aircraft or two up to starting off with a large number of aircraft on order or available or it may become a large operation by merging a number of smaller-scale operations. Despite the wide span of approaches the usual option is to start as a small operation, but one which is capable of being expanded if the operation is successful. Starting small and growing brings some governance problems with it, not the least of which is deciding how to establish a system of governance which is proportionate to the airline's operation and needs, but can be scaled up as the business grows. An additional problem is how to find some help in making the decisions on when and how to make changes to its governance.

Growth itself brings its own list of governance conundrums, as shown in this sample:

- How to communicate strategy and results effectively with an increasing number of staff.
- How to ensure that decision-making is appropriate to the size of the organisation with decisions made at the right level in the company, but understood by those who need to understand the decisions and those who will implement the decision.
- When to modify the reporting system for financial and operating information so that it remains effective and useable.
- When to introduce independent monitoring in the airline (e.g., internal audit or have committees to monitor some aspects of operations like operating safety).

All of these, and many more, problems need to be identified and a change made at the right time. To achieve this the governance practice and the airline's situation needs to be monitored so that the right change can be identified and a change made at the right time. A company's corporate governance structure is rather like a house which has to grow and be modified as the family and its needs grow.

The benefits

There are benefits to introducing formal corporate governance into a small airline or company. Most of these flow from the confidence and credibility that is generated when it becomes apparent that the company's founders and the Board of Directors are taking seriously the need to have openness and transparency in the way the airline is guided, monitored and managed. The fact that the board will run the business in a systematic way will carry a lot of weight with many of the stakeholders who are so essential during the early days of the company's life, such as staff, suppliers, financiers and customers.

Introducing good governance practices from the very start of the company should result in the airline having:

- A written strategy with measures and milestones and therefore Key Performance Indicators. This is really the business plan, together with a plan on how to communicate it.

- Clear reporting lines and lines of responsibility and authority, including a definition of the role of the Board of Directors with a formal delegation of certain powers to the company's senior management.
- An open process for making decisions together with the systematic delegation of authority.
- A process for identifying risks.
- A reporting system which monitors the operating and financial results of the business and contains a forecast for future results.

These are the absolute basics which the embryonic organisation needs.

The potential downsides of not dealing with the essentials of corporate governance matters are the reverse of the benefits, which is confusion in the way the company operates that will reduce the confidence of those indispensable initial stakeholders. The risks of not considering and implementing a basic, but complete governance system at the initial stage are substantial.

Starting small

Using the assumption that most airlines start their operation on a small scale, it is worthwhile discussing what the potential governance problems are and what the essential needs are for a start-up airline.

The possible governance problems a small operation may face include the following:

- If the founders of the company are closely involved in its operations, there is the likelihood that there will not be any clear differentiation between the individual founder's separate roles as, a shareholder, a director as well, potentially, as a manager; this latter conflict applies to every executive director of whatever size of company. This problem may be overcome by having an independent Chair of the board who can try to ensure that the separate roles are not allowed to overlap, particularly at board meetings; board meetings are for directors whereas Management Committee meetings are for management. Having a majority of INEDs on the board will also help because the INEDs will help ensure that all recommended decisions are fully and, potentially robustly, reviewed and in extremis they will be able to prevent an executive director from pushing an idea through the board. In additional, each director should have a description of their role on the board which is separate from their job description as manager of the airline. It is difficult to establish effective rules to prevent the potential problems which can arise in family owned/directed/managed companies because it is practically impossible to prohibit families from discussing the business when they meet socially.
- If all or a number of shareholders or directors are friends, there is the potential that decisions which should be made by the full Board of Directors at a

meeting where there is a quorum of directors, will actually be made when the group of friends meet informally away from the airline. This has the potential to become a real problem where some of the shareholders or directors are members of the same family. There is the danger that decisions will be made in the family home rather than in the board room.

There is some guidance on these two potential problems above and some warnings of possible business consequences which are published by Institutes of Directors, Institutes of Accountants as well as firms of lawyers and accountants. From a structural point of view, the best antidote is to have a majority of INEDs on the board and for the owners/directors to agree rules with the INEDs and non-executive directors to ensure that everyone is kept in the picture on off-site discussions and that any social exchanges on company business are repeated at a formal board meeting where a final decision can be made and minuted.

A number of decisions on corporate governance are forced onto the founders when they first form a company unless they decide to use the standard 'model' for the company's Articles of Association, but this is just another form of forced decision. They may wish to amend such items as:

- The maximum and minimum number of directors there will be on the board.
- The minimum number of times the board will meet every year.
- How the Chair of the board is to be elected.
- Whether directors are permitted to attend a board meeting by telephone or video.
- How the AGM will be run covering how the voting will be conducted and other routine matters.

It can be difficult for the founders and first directors to get guidance on the non-standard matters of governance although many of them are important. There are a great many ways that companies can organise themselves, so it is difficult for there to be one source reference material which covers every governance matter and the exact circumstances of the new company and its Board of Directors. The local Institute of Directors should be able to give some detailed guidance before the company is formed as should the lawyer forming the company. Some Institutes of Directors actually produce information suggesting how a company should organise and manage itself at various stages of its development, also when and how various governance matters should be introduced, and the advice may include the suggested contents of its financial and business reporting at each stage. Unfortunately, the information will be generalised because it has to be generally applicable to all companies in all industries. Hence, the founders and first directors will still have to form their own opinions on what to do and make some decisions.

It will be very useful if, as soon as the decision is made to go ahead with the airline project and before a company is formed, the potential directors of the company have been identified and they agree in principle to act. This gives time to establish how much knowledge and experience of governance each of the potential new directors has and for some training to be arranged for them to bring them up to speed. If some of the potential new directors have previous experience of companies and their formation, they may be able to give advice on the formation of the new company. Any required training may be given by attending courses run by local professional institutes, most frequently the local Institute of Directors. Also, many law firms and accounting firms produce pamphlets on aspects of governance particularly for smaller companies. There are many books on corporate governance and on how to achieve an efficient and effective board, and there are also sites on the internet which have information. These options provide a fairly inexpensive way to get the basic knowledge of corporate governance, so decisions can be made quickly as to how the airline will be governed. There are of course consultants who will be happy to advise the new company, but the suspicion is that the founders/directors will prefer to wait until the airline project has proved itself before considering engaging consultants.

Directors have responsibilities and are accountable for their actions. The limited liability of the company does not apply to directors, as they are personally liable if they are negligent or do not comply with the local company law. Virtually every company has a Directors and Officers insurance policy, often called 'D&O insurance' which provides insurance cover for a company's directors and managers. The cover is usually for claims arising from the decisions they have taken when acting as a director of the company. The policy covers the legal defence costs and any financial loss. It is important that a D&O policy is in place when directors are appointed.

The 'do it yourself' approach on how the new company will be governed has its advantages. It is easy to find lists of matters which need to be considered and decisions made, and it also leaves the founders/directors free to choose the approach they wish to follow. Despite the volume of information available, making the initial decisions on corporate governance can be difficult.

Early operations

During the early days of the new company's operation the Board of Directors will probably be required to meet frequently and regularly, partly because there will undoubtedly be a number of matters to discuss and this is the time when the way in which the board will operate will be formed, refined and formalised. This is also the time when the policy for communicating with staff should be in final form and implemented. This communication will lay the foundations for problem solving and building the management team.

It could be said that the early days of a new operation is when the directors really earn their fees. The subject of directors' fees during the early days of a new company can be a difficult problem. Having argued that the new airline's board should include some INEDs from the first day and that ideally INEDs should be the majority on the board, they will need to be paid for what they bring to the board room table, which is their experience, skills and contacts. The question is how to decide how much to pay them. The main options are to pay a fee or to sell the directors shares, or a combination of both. The INEDs are professionals and if paid should get a fee which is in line with the amount which would be paid to a professional lawyer, engineer or accountant: not necessarily the full hourly fee rate charged by a professional firm of lawyers etc., but the same rate that would be paid to a senior employee including an add-on for fringe benefits. If the INEDs only need to attend a fixed number of meetings each year then an annual fee can be calculated and offered, but if the commitment is open-ended it may be sensible to agree an hourly rate. In both circumstances the rate has to be 'in the market' (i.e., comparable to other companies in the same industry). The question of paying directors' fees to executive directors is more debatable, as being a director is certainly an extra responsibility on top of a management job, but it may be that the remuneration for the management job acknowledges the extra responsibilities of being a director. It will need to be borne in mind that, because the airline is newly formed, the reporting systems will be new and unproven, so the board and individual directors will not have the full level of support they need until the systems have settled down. This may mean that the directors need to spend more time than is normal to understand what is going on in the business. The estimated cost of directors' fees should be included in the airline's business plans.

It may be that the airline would prefer to offer an INED an allotment of its shares (i.e., a fixed number of shares and/or an option to buy shares in the future). It has been argued that all directors should own at least a minimum number of shares in the company, in the belief that this aligns the director's interest with that of the shareholders, and hence the board's decisions will always be made in the shareholders' best interests. Whether the argument is persuasive will depend on the board and ultimately the shareholders. If the airline's Board of Directors has decided to establish a Remuneration Committee, directors' fees and remuneration is a subject they should recommend on, although there may be a conflict of interest. If there is a conflict of interest the initial amount of directors' fees and the detail of their remuneration package should be decided by the founders. The area of directors' remuneration is a difficult one which can generate a lot of lively debate at AGMs.

Some professional directors' organisations recommend that at the first meeting of every Board of Directors of a new company, the following board committees should be formed:

- Audit Committee.
- Governance Committee.
- Risk Committee.

These are certain key board committees, but the board is likely to want to form all of the committees it will need to help it to do its work. In addition to those above, they may want to consider forming committees responsible for finance, safety and remuneration and any others they believe will be beneficial to their work.

The composition of the board during these early days will be very important. The need for a good balance of skills, experience and contacts appropriate to the airline's requirements was discussed in Chapter 2. The board of a company starting operations will need to have individuals who can make decisions quickly, have experience in the operations of smaller companies and experience or an appreciation of a range of operations. There will also be the need for at least one, preferably more, director who will keep their eye on the longer-term objectives of the company so that after the early operating period the airline's strategy can be reviewed and formally agreed. The actual composition of the first Board of Directors will be influenced by the strengths and weaknesses of the company's senior management.

Stable operations

Once the airline's operations have passed through the initial operating stage and the inevitable initial problems have resolved it will be the time for the Board of Directors to re-assess the company's strategy, hopefully confirming it, but if necessary, considering what changes should be made. This time may come after one, two or even three years of operations, but usually at the point when the board considers that the airline's operations are stable. A company's strategy is not an abstract idea, but a set of concrete achievable goals. With the strategy confirmed or modified the board can start to put in place more of the detail of the company's governance.

This post-initial stage in the airline's life is crucial because the decisions made and procedures set up at this point will form the basis for the airline's continued operation for the long-term. Hopefully in the early stages the board's Governance Committee will have been formed and operating, not necessarily issuing recommended complete documents and advice, but advising on how decisions are to be made and processes are agreed. It will also have been making notes of how the company has dealt with problems. If the committee has not been formed, one of the directors should have been made responsible for monitoring the airline's governance and advising on processes. Once the airline's operations are established and running as smoothly as any airline ever can, the committee should start the processes of:

- Finalising the Governance Committee's recommended instructions or manuals on its governance policies and practices which will include matters like, data protection, dealing with suppliers and tenders for suppliers. It will take some time for the recommended instructions to be formalised and a plan agreed for how they will be issued. When the manuals and instructions are issued the reasoning supporting the directions and guidelines should be explained to staff and any changes required to current systems highlighted

and the reason for the changes explained. It is important that any unacceptable practices or processes are eliminated quickly before they become part of the airline's 'this is how we do it' culture.
- Assessing how the board operates as a team and whether there is good participation from each individual director. Often this item falls within the remit of the Nomination Committee. It is not important which committee organises the assessment, the important part is that it is done and there is no duplication.
- Reviewing the information, financial and operating, regularly sent to the board, including its content, frequency and format. It may be that this review will be part of the Audit Committee's scope and the effort should not be duplicated. The terms of reference for each committee will have been approved by the board when the committees are first established.

At this stage it will also be useful for the Chair with the Nomination Committee to think through any change which is needed to the composition and balance of the board. It should be remembered that the board is an 'organisational tool', and exists to achieve a purpose. The skills, experience and contacts that a start-up airline needs on its board may not be the same as those needed by an airline which has a stable operation. If some changes are thought to be beneficial, an outline of the type of person or people the airline would like to have on its board should be prepared. Once the outline(s) have been agreed by the committee, the Chair and ultimately the board, a plan should be agreed to implement the change.

The Nomination Committee should also design an induction programme for the new director(s). This topic was discussed in Chapter 3. Most of the matters mentioned in the list of items to be included in the programme should be available, but the timing of the introduction of a new director may mean that some historical information (e.g., minutes of previous AGMs) may not be available.

Once all of the governance work, such as producing the airline's best practice manuals and policies, is in train and the board committees are established and have started their work, there is little on the governance front that is likely to change extensively, other than if there is some change to general recommendations on governance or there is a substantial change in the airline's business. The next significant point in the airline's history is likely to be when issuing additional shares in order to raise more capital for the business.

So far, the assumption has been that the initial operating period has shown that the airline's basic strategy has been successful and that although the airline's strategy may or may not need to be fine-tuned, it is essentially sound. If, however, the Board of Directors' assessment is that the airline has failed to prove that the current approach will produce a viable business, the directors will be faced with the challenge of recommending to the shareholders what to do next. The options are likely to be some form of orderly liquidation or selling the airline 'as is' (i.e., in its existing condition), or restructuring the operation. The governance issues involved in dealing with these options will be discussed later in the book.

Growing

It is likely that during its life a successful airline will periodically need to increase its equity, which is the amount of capital contributed by its shareholders. This is partly to ensure that there is a balance between the business risk being taken by the shareholders as well as by the institutions which are providing finance to assist the purchase of new assets. There are two basic ways to issue more shares; either by making a 'public offer', which is an offer to the general public to buy the shares, or by a 'private placement', which is an offer to one possible purchaser or to a short-list of potential investors to issue new shares to them. In either case the requirements of the procedure are broadly the same, with the major difference being that for a public offer the price of the company's share will be quoted on, and the shares traded on, a stock exchange. The stock exchange will have some minimum requirements which may influence the company's approach to corporate governance and the way it reports its financial and operating results to shareholders. Before making the final decision on whether to issue more shares and, if that is the best approach, whether to have a public offer or private placement, it is worthwhile for the Board of Directors to contact one or two experts in the speciality to get their advice and to ensure that everyone on the Board of Directors has a complete understanding of what is involved in both options and what the definite and possible effects on the company are likely to be. These experts are often known as corporate advisors or merchant bankers.

As part of the review as to whether to issue more shares, the directors should also consider the alternative of issuing bonds as this option could be a more attractive option for the company. A bond is more like a loan, where the purchaser advances money to the company by buying all or some of the bonds that it issues and in exchange receives a bond which is the company's undertaking to pay interest on agreed dates and to repay the amount advanced on an agreed date in the future. A bondholder does not have any ownership in the company and does not benefit from its performance, but will require that certain of the company's assets are pledged to the bondholders to ensure that interest is paid, and the principal is repaid, on the due dates. The procedures for issuing share or bonds is not dis-similar but the effects on the company can be quite different. The essential difference between issuing shares or issuing bonds is that, when more shares are issued the ownership of the company is held by more entities/ individuals and they all continue to share in the risks of the operation.

The actual and potential changes to the company's position flowing from issuing shares or bonds are probably sufficiently significant that the board should either establish a committee to examine the pros and cons and the effects in detail and to recommend a course of action or assign the same work to be done by the Audit or Finance Committees. Issuing shares or bonds is an important step and a considerable undertaking.

The timing of an issue of new shares or of bonds is very important as they need to be made when investors are keen to invest in the company and the industry. If the board concludes from the forecast of the airline's operating and financial position that either a share or bond issue will be needed in the foreseeable future, it is worthwhile starting the investigation into the options immediately, so that everything is clear and the foundation work done so the Board of Directors can move quickly when the timing is right.

At this point it is worth mentioning the different types of shares:

- 'Ordinary Shares', sometimes called 'Common Shares', which are the standard shares that participate in the risks of the company and receive any dividends declared; also if the company is liquidated, a share of the funds remaining after all creditors have been repaid.
- 'Preference Shares', which are shares having some form of preference. Most often this is to receive a minimum level of dividend each year before any distribution is made to holders of Ordinary Shares. The preference may also include some preference over Ordinary Shareholders in the event of the company's liquidation. Frequently theses shares do not have any voting rights and the company may have the right to buy them back from the holders at some stage.
- 'Founders Shares' are shares which may be issued to the founders of the company and often have some dividend restriction on them. It is not currently common to issue this type of share because they dilute the voting power of Ordinary Shareholders.

There is one other feature that needs to be taken into account when the board is thinking about possibly issuing more shares. It is not unusual for the first shareholders in a company to have 'pre-emption rights', sometimes called the 'right of first refusal', that is, the right for the current shareholders to buy any new shares issued by a company; this is a right not an obligation. If there is more than one shareholder, the right can be exercised by each shareholder individually, it is not necessary for all the shareholders to make the same decision.

Issuing new shares or bonds requires the approval of the company's shareholders. In most jurisdictions there are two ways to obtain shareholders' agreement. The first option is to ask shareholders at the company's AGM to approve a resolution authorising the Board of Directors to issue new shares up to a limit, (say) 10% of the number of shares currently, issued at their discretion. If this option is not used, an Extra-ordinary General Meeting (EGM) must be called so the shareholders can approve the issue of new shares. The shareholders also need to approve any issue of bonds.

Whichever form of transaction is used to increase the airline's funds, issuing shares by a public offer or through a private placement or by issuing bonds, the individuals or organisations paying over the money to the company will want detailed information on the company and what the new money will be used for. This information will be provided to potential investors (for shares) or lenders (for bonds) in a document issued by the company's Board of Directors called a 'prospectus'. The exact detail required to be included will vary depending on which financing option is used and also between jurisdictions, but essentially the contents will be:

- An update of the airline's business plan.
- Information on the results achieved to date.
- Details of how the new funds will be used in the business.
- The forecast effects on the airline's results of the new funds.

The sorts of detail which are required include:

- Details of the company, its name, registered office, operating locations, details of the directors, auditors, lawyers, bankers, a chart showing the management structure, a chart showing subsidiaries and associates.
- Corporate governance: details of the board committees, frequency the board and its committees meet, any conflicts of interest, a description of the responsibilities the Board of Directors has delegated to the CEO or Management Committee and how the board makes its decisions.
- Statement of the company's strategy (this need not include any information which is commercially sensitive), business opportunities available, business risks faced, its competitors.
- Summary of the company's assets analysed into types (e.g., aircraft, property).
- Summary of the company's loans and long-term obligations, such as leases.
- Statement of the company's financial results for the last two (in some cases five) years together with operating information.
- Dividend policy.
- Forecasts of the financial results, financial position and cash flow for the next five years, with an indication of longer-term prospects.
- List of obligations including any proposed or unsettled litigation, any significant supply contracts (e.g., for aircraft, simulators, advertising).

This is just a summary of the information for the prospectus. It is likely that the company's board will need help accumulating and presenting all the information needed and there are companies which can provide this service. Some of the information may need to be certified by experts (e.g., for the value of properties owned) and these certificates should also be included in the prospectus. Given the amount of information required, a wise board will start the work of preparing the prospectus well before the additional funds are actually needed in the business. All-in-all, raising large amounts of money is a substantial project in itself and there are significant costs involved.

It is clear from the summary of information required for a prospectus that good corporate governance principles need to be established and operating in the company as early as possible, ideally from the first day. Failure to have established good governance in the company will mean that it will be more difficult to convince potential investors or lenders that the company is well run.

Secondary capital expenditure

One of the governance problems facing the Board of Directors as an airline grows is how to deal with secondary capital expenditure. This is not a technical term, but a way to describe expenditure on capital assets which does not meet the normal criteria for 'capital expenditure'. Capital expenditure is money spent on an asset which will either increase the asset's useful life or increase the income earned by the asset. The board of an airline which has been operating for a few years will, almost inevitably, receive proposals to spend significant amounts on aircraft which will neither extend the economic life of the aircraft nor increase the income they can generate. The proposals will be for such things as more modern seats or improved inflight entertainment or changes to the mood lighting, and there may also be expenditure recommended for purely technical equipment like extra or updated weather radar. Each of these recommendations needs to be reviewed and a decision made by the board. Sometimes when a capital expenditure request is made 'the numbers will make the decision'; if there are funds available and more income will be generated by spending the money, then a return on investment can be calculated and probably a decision will be relatively easy to make. However, if there is not any extra income or reduction in cost, it becomes more difficult for the board to make a decision.

The problem facing the board is two-fold; whether the airline can afford to spend the money within the context of remaining solvent and separately whether the expense is justified. An up-to-date long-term cash flow forecast will help with the first problem, but the justification remains outstanding.

There is no easy solution for the Board of Directors to this type of problem. For capital expenditure on inflight customer amenities, the original proposal to purchase and fit out the aircraft should include estimates for replacing such items as seats when they are becoming worn or too expensive to maintain. It is difficult to include in the initial case for buying an aircraft, a forecast of when, or if, a significantly improved seat will be designed and become available. One approach to solving the problem is for each capital expenditure request for aircraft to include an allowance for 'upgrades' and for the board to put money aside each year to cover any upgrades. Any approved upgrade request will be funded from the money put aside. This possible solution again deals with the money side, but not necessarily with the justification. The sort of justification which will help the directors is one based on the airline's strategy. An example is, the strategy will, no doubt, have a section on 'Customer satisfaction' just as it will have one on 'Safety', and within these sections there will be measures or

KPIs with targets (e.g., 'Percentage of customers fully satisfied, target 80%'). If the actual customer satisfaction level is below 80% and the projected expenditure is expected and forecast to improve the actual satisfaction to the target level or perhaps above it, then the expenditure may be approved on that basis, because the investment has a measurable target to achieve.

Not one of these approaches is completely satisfactory, but the board has a responsibility to ensure that all the decisions it makes are made in the best interests of the company, short-term and long-term. In addition, they have the responsibility to ensure the company remains solvent.

Not according to plan...

All of the discussions in this chapter so far have assumed that the newly founded airline is successful. Its strategy has been proved to be workable, even if it has had to be modified in the light of experience, and that potentially the airline has a bright future. Sadly, this does not happen to every start-up airline and even well-established airlines can fail. Some fail simply because the airline's strategy is not attractive to customers, while others may have a good strategy, but one that is ahead of its time and the market is not yet ready for the approach or it may be that the market is already saturated with similar approaches.

The early warning of a potential problem may well come from the regular financial and operating forecasts produced by the airline's management for the Board of Directors. The early signs will be unfavourable changes in operating figures, revenue and passengers, and in efficiency measures, increasing cash deficits (where cash outflows are greater than the cash inflows), regular operating losses, consistent failure to meet revenue, cost and profit targets, deterioration in the solvency ratios mentioned in Chapter 5. A sign that the problem has arrived is difficulty in making routine payments to staff and creditors, but this point should not be reached if the directors read and interrogate the forecast and actual information regularly reported to them.

When the first signs of a problem come to light from the reports to the board, the Board of Directors will need to focus on examining the options open to them. Although the company will be currently solvent, if the figures indicate that this will change in the near-term, the board needs to give the highest priority to the airline's cash position and cash management because if insolvency arrives the directors' focus has to change to limiting the potential losses to creditors and to the protection of creditors.

The first reaction of most Boards of Directors is to initiate some change or series of changes which should correct and improve the situation bearing in mind that at this stage all decisions should be made for the benefit of the company. When this point is reached the directors will have to consider retaining external consultants to advise on changes which will remedy the situation. If this option concludes that a change will not ensure that the company will be successful or not sufficiently successful to correct the approaching problem, the directors will need to consider the future of the company.

At this stage there are probably still quite a lot of options to consider and the timing will be important. The urgency will depend on when insolvency is forecast to arrive. The best advice for the directors is that they take the best advice available as soon as they can. It is highly likely that a working group or committee of directors will be formed to analyse the impending problem and recommend possible solutions. Although all directors will be involved in the review and decision-making process of the options, the working group/committee will do the leg work.

There really are a large range of options for a company facing the problem of an approaching existential threat, but this depends on the jurisdiction. Some authorities permit companies to approach their creditors to seek to reduce the costs of services provided, and a classic example is to reduce lease charges. Other jurisdictions give the company a period of protection from the claims of creditors while the Board of Directors seeks to re-organise its business. Some of the measures only apply if insolvency is imminent, whereas other actions can be taken when the signs of a potential problem become more certain. It is important for the directors to get the best advice they can before they can decide which options to examine further. In many cases an orderly liquidation can minimise the loss to shareholders; 'liquidation' is the systematic disposal of assets, tangible and intangible, and using the proceeds to settle the amounts due to creditors; any remaining surplus will be paid to shareholders. An option favoured in some countries is for the board to appoint independent administrators to operate the company while disposing of its assets. The administrators aim at minimising the losses to the creditors while liquidating the company's assets.

Whichever course of action the Board of Directors conclude is the best for the creditors and shareholders needs to be put to the shareholders for approval. Once the agreement of the shareholders has been given, it may be necessary or advisable for the directors to meet major creditors to explain what action will be taken. This approach may forestall any legal action by the creditors against the company. Hopefully it is clear that this discussion of options is only for the unfortunate circumstances when the directors conclude that an airline will not be able to operate profitably in the foreseeable future. The courses of action discussed would not be appropriate for an airline which is just experiencing short-term trading difficulties. The Board of Directors has to make the crucial decisions on what the future of the airline is likely to be.

The principles of good governance do not really change, but how they are applied will probably change as the company goes through its various stages of development. The earlier that all of the recommended practices and processes involved in good corporate governance are introduced by the company, the better it will be for the company and its future.

7 Subsidiaries and related companies

When one company becomes a group of companies

It is not unusual for any developing company, and that includes an airline, to decide that certain of its operations, for example check-in services, would be most efficiently run through a separate company. Even an airline when it first starts may have some of its operations in separate companies, if its scale of operation warrants it or there are other good reasons. Once a company has formed a number of other companies, the collection of companies is usually called 'a group' which is a number of separate commercial firms with the same owner.

Having some operations in a separate company is not the same as sub-contracting those operations to an independent third party, as there are some basic differences between an airline sub-contracting an operation and establishing its own separate company to do the work. When sub-contracting, the airline does not participate in the profits or losses of the sub-contractor, but where the airline owns the contractor, as an investor, the contractor's profit or loss ultimately belongs to the airline.

In sub-contracting, the relationship is purchaser and seller; the supplier contracts with a purchaser, the airline, to provide certain defined services in exchange for a fee. The relationship between the airline and the supplier needs to be close and controlled. The challenge for the airline's managers is to find a practical way to manage the relationship with the supplier. When one company forms another company to provide services to it, the relationship between the two companies is also as purchaser and seller, but in addition to finding the right way to manage the operating relationship between the two companies, the Board of Directors of the airline (which is both a purchaser and an investor) needs to establish the best way to ensure that the supplier (seller) has complementary goals, business policies and corporate governance practices, despite the supplier being an independent company.

Reasons for having a separate company

There are many reasons why an airline might form a separate company. Local laws or business practices in the various locations in which the airline operates will or can have a significant influence as to whether having a separate company providing services is useful and helpful. The main reasons for considering whether there are overall advantages to the airline to have some operations in a separate company, fall into these general propositions:

- Local law or requirement: there may be a requirement that certain activities are conducted with a local partner, for example, opening an office to sell tickets or to service customers. There may also be other advantages to operating through a local company, for example, access to government grants.
- Projecting a local image: this may arise when an airline decides to put some or all of its ground operations in another country into a separate, locally incorporated, company.
- Part of risk management: performing some operations through separate companies may make it easier and less expensive to arrange insurance cover and to limit reputational risk.
- The possibility of locking into a lower cost base (e.g., rents, staff cost add-ons), by operating in another location or jurisdiction.
- Defining the tax position: most airlines operating internationally are covered by 'Avoidance of Double Taxation Agreements' which are international agreements exempting profits from the transport of passengers and cargo arising outside the airline's home country from being taxed by the other countries. This also means that profits from other, non-transport, activities (e.g., aircraft ground handling), are potentially taxable in other countries. The amount of profit to be taxed in a country can be more easily calculated if there is a separate local company.
- Forming a joint-venture with another company may result in better efficiency in some parts of the airline's operation.

It is important that a part of the process of an airline's Board of Directors' study into whether to establish a separate company includes a clear statement of what the new company's role will be, together with the advantages and disadvantages of separating the activity, because these will form the basis for directing the new company. If, for example, the proposal is to transfer check-in services into a separate company, then the advantages may be:

- Establishing a new business with its different risks and rewards.
- An ability to remunerate and organise staff appropriately for that part of the aviation industry rather than as part of an airline.
- Having systems which can deal with the likely level of training needed and the expected higher rate of staff turnover.
- More flexible systems increasing the ability to offer services to other airlines.
- Establishing a local identity.
- Establishing a potentially profitable business which may be able to be developed further.

The possible disadvantages might include:

- Resistance from existing staff to transfer to the new company together with some anxiety amongst the remaining staff.
- The costs and time required to establish the company and its reputation.
- The possibility that the level of service to the parent company may be reduced when there is a conflict with the schedules of other airlines.
- Increasing the business risk to the parent because any problem in serving another airline may become attached to the parent company's reputation.
- The possible diversion of the airline's management effort away from running the airline, if the new company is not profitable or has some other substantial difficulty.

It will be noted that the possibility of profit or loss is only one of the considerations. Before making a decision, the board will have to consider the proposal in the light of the airline's own agreed strategies.

Types of entities

Before mentioning the most common types of company used within a group of companies, it is worthwhile mentioning that one of the simplest ways to separate part of an operation is for the company to establish a 'branch' or office in a separate location with its own operation and with its own set of accounts. Establishing a branch tends to be a simple process which is an advantage and it can be used to project a local image. Generally, the main company remains responsible for the actions and liabilities of the branch, so there is not any separation of liabilities or risk. It can be more difficult to calculate the profit arising in the branch for local taxation purposes. This type of structure is not popular for substantial operations.

Having dismissed the 'branch' it is appropriate to look at the most common form of structures used within a group of companies which are subsidiaries, associates and investments.

Subsidiary

A subsidiary company is a company which is either owned and controlled by one company, in which case it is a 'wholly owned subsidiary' or where it is owned by more than one company with one company owning a controlling interest in it, it is simply a 'subsidiary'. The company with the controlling interest is usually referred to as the 'parent company' or 'holding company'. A controlling interest in normal circumstances means that the parent company has at least the number of shares which ensures that it can control the result of all board and shareholder resolutions. In most jurisdictions this means the parent must own at least 50% plus one of the shares. In jurisdictions where a two-thirds majority is needed to pass a resolution, the parent will need at least two-thirds of the shares to have control. Despite the parent having the controlling interest, the subsidiary remains a separate legal entity, independent from the parent company and in general with its own Board of Directors with the same responsibilities and duties as any other board. In some jurisdictions there is an exception to this basic rule which permits the board of a wholly owned subsidiary, if required by the parent, to make decisions which benefit the parent at the expense of the subsidiary, but this tends to be the exception rather than the rule.

The potential conflict between the control which the parent can exercise through its voting rights at board and shareholder meetings and the subsidiary's directors' responsibility to make decisions in the best interests of the subsidiary, can be a concern to directors both on the parent's board and the subsidiary's board. The potential for conflict can be resolved by including in the subsidiary's Articles of Association:

- A clause defining and limiting the scope of business the subsidiary can engage in; the clause will list what business activities the subsidiary may operate or may not operate or both. In effect this clause or clauses will become the subsidiary's strategies.
- Restrictive clauses requiring the subsidiary's Board of Directors to get shareholder approval for certain important matters, for example, for committing to buying or selling assets of more than a specified value, or arranging any borrowing, whether secured or unsecured, of any amount in excess of a certain limit. In addition, requiring shareholders' approval for all operating and financial plans including those for the following year. The shareholders may add other matters as they think appropriate. It should be remembered that the directors of the subsidiary must present the actual operating and financial results to shareholders at the AGM in any event and to answer shareholders' questions.

A further protection may be to appoint a director of the parent company to the board of the subsidiary, although if this happens the nominated director is still required to make decisions that will benefit the subsidiary when they are at the subsidiary company's board meeting. The nominated director must declare their conflict of interest, but will be able to flag to the subsidiary's board when any matter arises on which the shareholders should be briefed before a decision is made.

Where the subsidiary is a joint-venture (i.e., not a wholly owned subsidiary), the minority shareholder or shareholders will probably want similar restrictions in the subsidiary's Articles of Association to try and ensure that they are fully informed about what is going on as well as for their protection. There are laws which protect the interests of minority shareholders, but they can be difficult to enforce.

Within the framework and restrictions detailed in the clauses of the subsidiary's Article of Association, the board of the subsidiary is able to make decisions in the same way as any other Board of Directors.

Associate

The other common form of company is the 'associate' or 'associated' company. The definition of an associated company can be rather complicated. The easy part of the definition is that a company is an associated company of the shareholder if it is not a subsidiary and the shareholder has a substantial interest, usually at least 20% of the voting rights, plans to be a long-term shareholder and can exercise significant influence on the company's operation. Determining whether the shareholders have a significant influence can be difficult. Significant influence usually involves the shareholder:

- Having the ability to appoint members to the company's Board of Directors.
- Providing managers and/or crucial technical information to the company.
- Taking part in the development and decisions on policy, including such matters as dividends.
- Having frequent, significant business transactions with the company.

The associated company may have similar restrictions in its Articles of Association as a subsidiary. Hence, the shareholders will have the ability to discuss matters which are important to them before the final decision is made by the Board of Directors of the subsidiary or associate.

The accounting treatment for subsidiaries and associated companies is different and is generally covered by International Financial Reporting Standards. A list of the company's main subsidiaries and associates will be found in its annual report.

Any transactions between a shareholder and its subsidiaries and associates, must be 'at arm's length'. This concept is important. A transaction 'at arm's length' is one which does not take into account the relationship between a company's shareholders and the company and is the transaction as would be agreed between two independent parties on the basis of market rates without any pressure on either party to conclude the transaction, which is the principle of 'willing buyer and willing seller'. This approach must apply not only to any purchases or sales of assets, but any charges made for goods and services provided. Failure to adopt this approach:

- Will distort the financial results of both parties.
- Will probably make it difficult for the shareholders to evaluate the efficiency of the subsidiary or associate.
- May defraud other shareholders.
- Is a conflict of interest.
- Can lead to taxation problems.

Investment

Any other shareholding which the company may have, which is not a subsidiary or an associate, will be classified as an 'investment' and reported separately in the company's annual report. In most cases investments are held simply for the income they generate or the possibility of an increase in value, whereas the investment in subsidiaries and associates should mean that the company's overall operations are more efficient and/or more profitable. Given an airline's thirst for money it is probable that an airline will not have a significant number of pure investments.

There may, however, be a separate vehicle, possibly a trust or company, holding the investments of the airline's pension scheme which will most likely be run by trustees and the portfolio managed by professional investment managers. Isolating the assets held for income and to meet the pension entitlements in an entity separate from the airline and its operating assets gives a measure of security to currently employed staff who will benefit from a pension and to existing pensioners and it is also a part of good governance practice. In many jurisdictions it is a legal requirement for the assets of the company's pension scheme to be held separately from the operating business. There have been examples where a company has not isolated the pension scheme's investments and when the company had trading difficulties there was not sufficient income or saleable assets to meet the company's liabilities to its pensioners.

Governance in subsidiaries, associates and investments

Good corporate governance is as important in subsidiaries, associates and investments as it is to the investor. Some investors argue that no investment should be made in another company unless good governance practices have already been established, but following this line could mean that an investor misses an opportunity to co-operate in an operation which has the potential to benefit the investor's operations and financial results. A more constructive approach is to arrange for the subsidiaries and associates to develop a complete set of governance policies and practices together with the means to monitor them. The development of good governance, which includes business ethics for the whole organisation, builds trust not only amongst shareholders because operations are open and transparent, but also with staff, suppliers and general stakeholders. This trust and confidence, which are at the core of its reputation, are a valuable asset for the company.

It should be relatively simple to ensure that a wholly owned subsidiary has a workable governance set-up, because the parent company will already have done the ground work for itself, and the subsidiary can adopt the same approach, policies and practices. This may not be the complete solution because the nature of the subsidiary's operations and size of the management team may mean that the subsidiary is not able to duplicate and operate the parent company's system. It will be recalled from Chapter 6 that governance practices are likely to change as the size of the company's operation changes and that small-scale operations tend to meet the principles of good governance in a different way to a large publicly quoted company. A further consideration for the subsidiary is the normal set of practices in the part of the industry they operate in. It may be that the parent pays creditors in terms of taking 30 days' credit, while in the subsidiary's area of business the accepted practice may be for a shorter or longer period. The parent's policies and practices may need some modification, but in any event the parent's experience and advice will be of great help to the subsidiary. The parent company will wish to know the differences between its governance system and the subsidiary's. It is likely that the parent company's internal audit team will periodically want to review the operation of the wholly owned subsidiary's corporate governance and the reasons for the differences from the parent's approach. Depending on the size of the subsidiary, it may be beneficial for a representative from the subsidiary to join the parent company's Governance Committee. If the subsidiary has its own Governance Committee, a copy of its minutes will be sent to the parent's Governance Committee. This practice of each committee having representatives from subsidiaries helps communication and education, but care must be taken that a committee does not become so large that it does not operate effectively. The parent should receive copies of the minutes of the subsidiary's board and board committee meetings.

Dealing with good governance in those subsidiaries, which are not wholly owned, and in associates may need some more work and discussion because there may be other shareholders whose views and priorities should be taken into account. Not all the shareholders may be at the same stage of development of their governance processes and standards and this may be because the operations of the other shareholders are on a larger or smaller scale than the investor's or because there is a difference in their approach to governance. Any differences in approach need to be discussed in detail before the investment is made. It should be possible to find some common ground and agree a shared view of what to do, but in the worst case, if the companies' views are irreconcilable, it may be advisable that the investment does not take place. Corporate governance is an issue for discussion before the investment is agreed or before a new shareholder invests in the company, because the governance practices and standards of subsidiaries can reflect badly on the shareholders if a significant problem develops. Ways to monitor the operation of the governance processes and standards in the subsidiary also need to be agreed, as well as how the shareholders will be kept informed of the subsidiary's operations. Although there are potential difficulties in developing good governance within subsidiaries and associates, there is also the potential that an investor can learn from the techniques and approaches used by some of the other investors, or from the policies and procedures the subsidiary adopts. No entity has the monopoly on good governance and because it is a subject which is alive, it must be kept under critical review; it is very useful to keep an open mind.

The essentials of the governance of a subsidiary, which is not wholly owned nor a joint-venture in which one shareholder owns the majority of its shares, will be found in the agreements made between the shareholders before the subsidiary or joint-venture is formed. If an investor is buying into an existing company, the governance organisation and practices will already be established, even if they have not yet been formalised or are reflected in manuals. Potential investors must be comfortable with the company's approach to its governance. If the company has yet to be formed there is the chance for the potential shareholders to agree a governance structure and practices which will suit them all. Even if there is a difference in the size of the shareholdings between shareholders of the new company, the governance details need to be agreed and accepted by all as if amongst equals, not forced by one shareholder.

There are potential difficulties which may arise when discussing and agreeing the new company's governance. If there are shareholders from different jurisdictions and cultures, those negotiating the governance agreement will need to understand these differences. The sorts of differences in background which may surface during the pre-investment discussions are as follows:

- There may be different approaches to involving staff in decision-making. In some countries it is usual to have staff representatives on the board, in others on a Management Board, and in some, the board simply explains decisions to the staff or their representatives.
- In some jurisdictions it is common to have bicameral boards while in others a unitary board is used; this was discussed in Chapter 1.
- There may be differences in the business cultures which can arise, for example, understanding what a contract represents, a binding agreement or an agreement reflecting an intention. Other examples are the degree to which a joint-venture is separated from its investors, or the responsibility for staff transferred from a shareholder to the new company.

The parties negotiating the governance agreement need to be aware of the differences and backgrounds and to be sensitive to them.

Although in many cases a new company will be formed to operate a joint-venture, the joint-venture may not always be incorporated into a separate company. It may be that the joint-venture is some form of co-operation agreement between two companies (e.g., in developing a new seat) or one company selling its product through another company's sales network. If a new company is to be established, the final agreement reached between all the parties in the joint-venture will be evidenced in the 'joint-venture agreement(s)'. This is an agreement or set of agreements between two or more parties to form an entity to carry out a business. The essentials of the understanding between the parties will include:

- A statement of the new company's strategies together with milestones for implementing it and Key Performance Indicators for measuring progress.

- The process for reviewing and amending the strategies.
- The contribution each shareholder will make to the new company, whether money, advice, management, technical staff or any combination of these.

Somewhere in the agreements will be a section dealing with the governance of the new company. The important areas to discuss and agree are:

- The structure and composition of the board, covering the skills, experience and contacts needed, and the number of directors, executive, non-executive and INEDs. Usually some or all of the shareholders will nominate a director to its board as an executive or non-executive director. This means the nominated director will have a potential conflict of interest. If the nominated individual is also a director of one of the shareholders, the potential for conflicts of interest increases. There is also the chance that the nominated director will be seen as a 'spy' for the investor. The board of the company should include at least one INED and one INED should have had experience in the potential governance problems which can arise in joint-ventures. The INED could be the Chair of the board and this may be the most useful position if there is to be only one INED on the board. Including an INED may make the board seem to be too large, but the benefits of independence and a view from the outside should outweigh any inconvenience. In addition to introducing the important element of having an independent view on board matters, the INED may be able to give advice, if and when there is a difference of opinion amongst a shareholder-nominated director and the other directors.
- The board committees which are to be established. It will also be important to agree how the committees will communicate, and possibly co-operate, with the shareholders. Each committee will need to be dealt with separately. The communication and co-operation between all the investors on Risk Committees matters will be different from the matters considered by the Nomination Committee. If the function of a committee is carried out by a director because of the new company's small scale, then the method of communication and co-operation should be described and minuted, so the directors are clear about what is expected.
- How the board as a team and as individual directors will be assessed and how frequently.
- How voting at board meetings and the AGM will be conducted and what size of majority is required to pass a resolution. This can be a complicated issue if the joint-venture has two or more investors which are from different jurisdictions.
- Who the company's external auditors, bankers and lawyers will be.
- Any limits on the authority of the Board of Directors in such areas as buying or selling assets and arranging loans. Usually a maximum monetary limit will be set for the board with all other decisions going to the shareholders for decision.

- Any limits on the matters which the board can delegate to the company's operating management. There may even be a list of the items which must be delegated to management.
- Ways in which the board will communicate with shareholders, either individually or as a group, matters like reviewing short-term and long-term plans. This is an important area to agree and document, partly because effective communication should ensure that the investors have the information they need, but also it should mean that all investors will have sufficient confidence in the board and management of the company to leave them alone to run the company within the parameters the shareholders have agreed.
- If possible, to establish a framework for a manual of Business Ethics. If this cannot be done in advance, it should be high on the agenda for the new company's board.
- Standards to be used for the statutory financial and operating report, particularly on environmental, social and governance matters, together with any plan to increase the span and improve the degree of reporting.
- A comment on any concerns which any of the investors may have about ensuring there is equality of treatment between investors. This can be important where there is a mixture of large and small investors.

The board of each of the shareholders will also need to make some decisions relating to the new company and these apply equally to wholly owned subsidiaries, subsidiaries, associates and investments. An essential step is for each shareholder to establish the risks involved in operating part of the business through a wholly owned subsidiary or another company, together with any risks which are unique to an individual company. Being involved with others can mean an increased risk to each shareholder's reputation, for example a strike in a joint-venture company supplying services to the airline can reflect on the airline itself. The investor's board must also make time to consider and respond to the information from the wholly owned subsidiary or joint-venture. Before making the decision to invest, the board will have established how and at what times the board will assess the subsidiary's or investment's performance as well as its contribution to achieving the airline's strategies.

Investments

The ability of shareholders to influence the governance policies and practices of investments is limited, but investors are not without some power. The latest annual report issued by the potential investment will include a description of the governance in the company as well as factual information on such things as, the number of board meetings each director attended, who is on each of the board committees and the remuneration of each director and senior management. If there are any areas where clarification is needed the investor can write to the company requesting an explanation and/or ask questions at the company's AGM.

The issue of the remuneration of directors, executive directors and senior management has become sufficiently important that guidelines have been developed in some jurisdictions which give shareholders the right to receive a specific report giving details and comparisons, to interrogate the directors on the report and vote either to accept or reject it. Without becoming involved in the debate on directors' and executive remuneration, it is worth noting that one of the reasons for introducing remunerations reports and shareholders potentially voting on them, is because shareholders have been unhappy about the way companies have responded to shareholders' questions and concerns in the past. It illustrates that governance standards are being set by pressure from society and that all aspects of good governance are of interest to society. All serious concerns from shareholders and stakeholders should be heard by the Board of Directors and a response prepared. Failure to respond can lead to future problems, possibly even legislation in the future and a loss of confidence by investors.

Suppliers

There is pressure on organisations to try and ensure that a company's governance practices and ethical values are followed by all of its suppliers. This is particularly the case if the company which places the order has a public brand and reputation, and will include most, if not all, airlines. There are many examples where a company has suffered both financial loss and the loss of public reputation because one of its suppliers has not followed the buyer's own standards of governance. The supplier has been seen as an extension of the buyer and, in the worst cases, the argument has been made that the buyer is trying to circumvent its own governance standards and practices by buying in goods and services. Where reputation is concerned allegations can do as much damage as proven facts. It follows therefore that a company should seek to agree with all its suppliers of goods or services, both nationally and internationally, that they will follow the buyer's governance and ethics standards and practices. This will be a protection for the buyer's reputation, and should continue the process of raising governance standards in the commercial world. Further, it is an extension of the assertion that good governance benefits businesses as well as society.

Initially, extending governance to suppliers sounds like an easy idea to implement, a 'win–win' situation (a situation where every party benefits), but there can be problems. If these problems cannot be resolved after discussions, the supplier's governance and ethics practices may remain a substantial business risk for the buyer. Imagine the case of a small airline, wishing to buy a new design of passenger seat with particular features the airline wants. If, after a review, the airline concludes that the supplier's governance standards could and should be improved, but the supplier declines, the airline's options are stark. In their simplest form the options are, to accept the supplier's resistance and try to manage the business risk while ordering the seats, or ordering less-desirable seats from another supplier with acceptable standards or one who is willing to change their standards. This is not an easy business decision for the Board of Directors to make. A similar situation arises in an international transaction if the supplier's standards are considered to be above the average in their own country or their practices are considered normal, acceptable and legal business practices in the supplier's country. There have been cases where companies have simply refused to buy from suppliers who do not agree to follow their standards and usually, they have been very large companies. The possible situation discussed in this paragraph may be transitional and that in the future universal standards will be adopted, but to date the application of corporate governance best practices has been steady, but nevertheless slow, so the business risk to small companies is likely to continue to exist for some time yet. It is difficult in all cases to argue that 'virtue is its own reward' (i.e., doing something solely because it is ethically correct, is more important than doing it to receive some identifiable benefit).

When discussing governance and ethics issues, generally the major contentious issues between the buyer and supplier centre around:

- Arranging safe working and healthy conditions for all staff.
- Paying all staff a 'living wage'. This term is difficult to define, but the definition is usually something like, sufficient to enable each staff member to afford food, shelter and health treatment. Even with an agreed definition, it is often difficult to establish what the appropriate numbers are.

These two items are often wrapped together in the requirement to comply with the local law to pay a minimum wage and to provide a safe working environment. In addition, there may be discussions around:

- Establishing a complaints procedure so that all staff may voice or seek an explanation on any grievance they may have, without the possibility of punishment or discrimination. There may well be cultural differences to be examined for this to work.
- Avoiding conflicts of interest including the issue of giving or soliciting or accepting gifts or advantages. This area also needs to take into account normal business practices in each country, where the giving and receiving of gifts is normal practice, or circumstances when it is and is not acceptable, together with limits. It is helpful if it can be agreed that all gifts received and made are recorded.
- Minimising the unfavourable environmental effects of the supplier's processes. This can also be a challenging discussion particularly in a developing country. It is unlikely that a solution will be found in the short-term and it may require the buyer to assist the supplier in such areas as specifying more natural materials or different processes.

Another problematic issue is very often how the supplier's compliance with the agreed governance and ethics standards will be monitored, even though in theory, the local national government is responsible for ensuring compliance with the law on such matters as minimum wages and safe working conditions.

A large organisation may be able to use its own internal audit staff to periodically review the supplier's compliance or more likely engage a local independent party to do the work, but a small organisation may not have the resources or be able to afford the cost of monitoring compliance. One potential solution is for smaller buyers to have access to the results of the reviews made by larger buyers and making a, hopefully, small contribution to the cost.

It is becoming a usual practice for a company to include in its statutory annual report an explanation of how it manages and monitors the governance of all the companies in which it has an interest.

One of the results of all this hard work on governance may be an increase in the buyer's costs and against that extra cost must offset the benefit of a reduction in the buyer's business risk. As with so many aspects of good governance, it is easy to calculate the cost, but very difficult to put a value on the benefits.

A company's Board of Directors is responsible for ensuring that the company has good and relevant governance standards and practices. More than that, it is responsible for influencing, to the greatest extent possible, the standards and practices in all of the organisations it has a direct interest in, as well as the companies it deals with, to ensure that the standards and practices meet or exceed current best practice. The Board of Directors needs to allocate time in its board work plan to do this. Good governance does not stop at the board room door, either side, going in or coming out.

8 Reviews

Corporate governance must be kept up to date and relevant

The Board of Directors of any company is responsible for the governance of that company. Naturally this involves ensuring that there is a process in place and in use for making sure that the company's governance, policies and practices are up to date. Being up to date means not just reflecting the current policies and practices that are being used in the company, but also incorporating those new and recommended policies and practices which are appropriate to the company.

It may be that the wording of the original corporate governance guidelines needs to be clarified or expanded. Perhaps some of the processes used in the company have been changed and if the new process is an acceptable substitute, the new process must be documented and approved by the board. It could be that the company's operations have changed in scale and the original approach to such things as the authority levels for approving expenditure no longer work efficiently. Change is constant.

Keeping policies up to date can be a substantial and significant job, particularly in a company which is growing, and it is only one of the responsibilities which the Board of Directors has, so in most organisations the directors have to balance their monitoring of governance practices and approving changes with all their other responsibilities. The board should use all the avenues of assistance that are available. The main sources of support are, to use the services of independent advisors or to arrange to have internal resources qualified to review governance practices or to use a combination of the two. Most companies choose the third option using both the services of the company's external auditors and its Internal Audit Department because both of these organisations have the knowledge of how the company's policies and practices are actually working. The information gained during their routine work enables them to compare actual practice in the company with those approved policies and practices. In addition, they will have a view as to whether there are any gaps in the policies, practices or working practices. Also, both organisations should be up to date with governance best practices.

During their review of the company's statutory financial statements the external auditors perform tasks which relate to the company's governance. The company's Internal Audit Department can be charged with commenting on how well the currently agreed governance practices are working during the course of their review work and therefore will be able to recommend whether current practices should be changed or the current requirements reinforced. The board should understand the work of the external and internal auditors so they can derive the maximum benefit from it.

External audit

The company's directors are solely responsible for the preparation of the company's statutory accounts which involves deciding on the company's accounting principles, standards and practices. In addition, they are responsible for ensuring the statutory accounts (i.e., the ones included in the reports to shareholders) show a true and fair view of the company's financial result and financial position. The directors are also responsible for introducing effective internal controls and practices to ensure the financial statements are free from any material error or omission.

The external auditors, who are an independent firm of accountants authorised to examine financial statements, are responsible for reviewing and assessing those decisions of the Board of Directors which are reflected in the financial statements. Consequently, it is useful for the board to understand what work the external auditors do in relation to both the company's governance and the work of the board. The main function of the company's external auditors is to form an opinion on the accuracy of the company's statutory reports, the Profit and Loss Account, the Statement of Financial Position and the Cash Flow Statement and then to communicate their opinion to the company's shareholders. The auditors want to be sure, within reasonable limits, that the financial statements are free from any 'material' misstatement arising from fraud or error.

'Material' is not an absolute amount of money, but an amount which is relatively important within the context of the individual company. Expenditure of USD10,000 may not be material to a company which has annual costs of USD10,000,000,000, but will probably be material to one with expenses of USD100,000. Within the accounting profession there are definitions and formulae for calculating materiality. A working definition of material is, 'any amount which, if misstated, omitted or obscured in the company's statutory reports, would be likely to influence the decision-making of any person entitled to use the information'. There are many formulae for calculating materiality and these tend to take into account the amount of the company's equity, total assets, revenue, operating profit and profit before and after taxation. Directors must bear the auditor's approach on materiality in mind when writing their annual report to shareholders.

Being reasonably sure really means having a high level of confidence, but this does not guarantee that there is not some material misstatement in the statements, especially if there is some form of fraud or deliberate mis-accounting.

It is important to note that the primary function of the external audit is not to reveal fraud, but to assess the accuracy of the figures presented to the shareholders. The external auditors may find a fraud during their work and they will, naturally, advise the details to the Board of Directors for their action. The external auditors are employed to report on the statutory reports to the shareholders, but they tend to be appointed on the recommendation of the directors and the shareholders usually authorise the directors to agree the auditor's fees for their work.

Key areas

The external auditors' responsibilities include, reviewing the company's significant accounting policies and forming an opinion as to whether they are appropriate, examining the company's reporting systems and deciding whether they are appropriate, also whether the directors' estimates included in the statutory accounts are reasonable. The work of the external auditors will be governed by the Auditing Standards issued by the local accounting institute or the local government. Although the external auditors will make test checks on all of the company's accounting systems, they focus mainly on key matters which can have a significant effect on the statutory accounts. In their formal report expressing their opinion on the accounts which is included with the accounts in the company's annual report, the auditors will detail why each matter they have reviewed is key and what work they have done during the review of each matter. Naturally the key matters vary industry by industry and possibly company by company, but for an airline they will usually include the following:

- When revenue is reported as being earnt; most passenger and cargo revenues are received in advance and recorded in an account often named something like 'Revenue Received in Advance'. The auditors are interested in the basis used to decide when the revenue is transferred from the Revenue Received in Advance account to the airline's Profit and Loss Account, as well as the system used to make the transfer. The principle used is that the revenue is transferred when it has been 'earnt', that is, when the passenger has flown and the cargo has been shipped. Any complication arises from such factors as, passenger journeys involving several flights spread over weeks or even months with some flights on other airlines. The revenue accounting system must track each flight so that revenue is transferred after each flight. In addition, the system must control the charges from other airlines for flights which the passenger took with them. The passenger revenue transfers are further complicated if the airline has a mileage reward programme. The accounting for revenue can be much simpler for airlines which only issue tickets for their own flights, do not accept bookings from or to other airlines and do not have a mileage programme.
- If the airline has entered into financial contracts with third parties to manage its exposure to changes in fuel prices, interest rates or exchange rates (these

are often called 'derivatives') the system used to agree, record and periodically revalue these transactions will be examined by the auditors. They will also assess the effectiveness of the company's financial management. This review is a very useful addition to the work of the board's Finance Committee and Audit Committee and gives the board some comfort that these risks are being systematically managed.
- An airline's taxation position can be complicated and disputes with authorities may well arise. In addition, the airline may be engaged in litigation or disputes with suppliers, regulatory authorities or other third parties. The board is required to include in the company's statutory accounts figures for the possible financial effect of these disputes once they have been finalised. In practice it is likely that the airline's management will already have given the board a report detailing the dispute which will have been brought up to date periodically, commenting on the progress of the dispute and the possible financial and other results. The board must decide on the commentary and figures, if any, to be included in the statutory accounts. The auditors will review both the assessment and the directors' decisions.
- The value of aircraft and their equipment have a major effect on the financial result in airline's statutory accounts through the charge for depreciation (Profit and Loss Account) and the calculation of the cost and value in the Balance Sheet (Financial Position). The auditors will examine, and may challenge, all the elements in the calculation. These are, the basis agreed by the directors of what cost items are to be included in the cost of aircraft, the estimate for the useful life of each aircraft in the airline's fleet, the estimated value of each aircraft on disposal. When these elements are determined the auditors will review the basis to be used to charge the reduction in value (cost less disposal value) against the revenue earned every year. The auditors may contact third parties to cross-check valuations and also review how accurate the directors' past estimates have been. The auditors' independent review is helpful to the board's work.
- Aircraft maintenance is a significant cost for an airline whether the aircraft is owned or leased. The airline will have its own maintenance standards and procedures, but may have to follow different obligations under any lease contracts it has. Aircraft have a series of mandatory maintenance checks as well as routine and emergency maintenance. It is usual for an airline to estimate the probable cost of each of the mandatory checks, the approximate timing and cost of the next check or checks: this amount will be charged to the airline's operating costs on an acceptable basis possibly on the estimated utilisation of the aircraft. The auditors will review all the forecasts and compare them with actual costs and timing of past estimates to assess whether the airline's forecast is reasonable. The airline's systems for tracking actual maintenance costs will also be reviewed.
- Review the airline's internal control systems and procedures to judge whether they are appropriate, adequate and likely to detect any material misstatement or omission. The purpose of this review is to establish the

reasonableness of the financial statements rather than to find fraud. In addition, the auditors will form a view on whether the airline's accounting policies and approach to estimating factors relating to costs (e.g., depreciation) are appropriate.
- When the local laws or regulations require the directors to prepare a business review (i.e., a discussion of the financial figures reported together with an assessment of the business as a whole), the auditor may express a view on the completeness of the business review.

Tables 8.1 and 8.2 illustrate the relative importance in a fictitious airline of the various account headings used in the statutory reports when reporting the airline's financial result and financial position.

Table 8.1 Simplified summary airline operating account

		%
Revenue	Passenger and Cargo Revenue	91
	Other Services	9
TOTAL		100
Expenses	Fuel	32
	Staff	18
	Route and Parking	16
	Depreciation	15
	Aircraft Maintenance	9
	Inflight Services	5
	Other	5
TOTAL		100

Note: The percentages have been rounded.

Table 8.2 Simplified airline assets and liabilities

		%
Assets	Aircraft and Property	62
	Investments in Associates	14
	Liquid Funds	8
	Debtors	7
	Intangibles	5
	Other Investments	3
	Stocks	1
TOTAL		100
Liabilities	Loans and Leases	65
	Creditors	14
	Income Received in Advance	11
	Deferred Tax	10
TOTAL		100

Note: The percentages have been rounded.

Advice to the board and its committees

The examples above clearly show that although the external auditors' work is directed at providing an opinion to the company's shareholders, a lot of the work they do can also give the Board of Directors some comfort about the accuracy of the routine information presented to them. That routine information will be used as the basis for decision-making, monitoring and directing the airline's financial result and financial position, and also when reviewing the forecasts of financial results, financial and cash positions. It is quite usual for the auditors to give the board a 'management letter' in which they comment on the efficiency, accuracy and timeliness of the company's accounting systems and they will include recommendations on improvements. This is useful feedback for the board in general and its Audit, Finance and Risk Committees.

The external auditors' closest relationship in the airline is likely to be with the Audit Committee. Both will work together to finalise the scope of their review and the timing of their audit. Once the audit has been completed, the auditors will discuss the contents of their management letter and hear from the committee what action they propose to take. It is likely that the detailed follow-up work will be done by the airline's Internal Audit Department.

Making changes

The knowledge which the external auditors gain during the course of their annual review and evaluation of the airline's financial systems and procedures can be extremely useful when the airline considers any change to its systems or procedures. In the past it has not been unusual for the auditors to be employed as consultants to review, assess and recommend changes to a company's accounting systems and then to assist in the implementation of the changes, but this practice is now not encouraged. Involving the external auditors too closely in designing and implementing changes to the airline's financial systems and procedures produces the concern that they may compromise their ability to objectively review and assess the financial figures in the statutory accounts as well as the systems and assumptions used to produce the figures. This would reduce the value of the opinion they give on the financial statutory reports to shareholders. Having the auditors design, implement then express an opinion on the same accounting system raises the classic question, 'Quis custodiet ipsos custodes?' freely translated as 'Who will watch the watchmen?' The external auditors' primary function is to express an opinion on the airline's financial figures and they must avoid any suggestion of a conflict of interest as this may compromise the value of the opinion they give on the financial part of the statutory report, in the eyes of not only the shareholders but also of the general public. Nevertheless, the auditors can provide useful background and constructive suggestions to whoever is retained to review and possibly change the accounting systems.

Internal audit

The Board of Directors needs some assurance that the details of the company's corporate governance are working and working well and that the procedures are still appropriate for the company's current operations. Achieving this requires reviews to be made of the detail of the procedures as they are used every day in the airline's operation. The work of the airline's external auditors can be very useful to the Board of Directors and their quest to ensure their system of governance is up to date, relevant and effective. However, because the external auditors focus their attention on expressing an opinion on the reasonableness of the airline's financial statements sent to the shareholders, their review of the airline's accounting systems is not completely comprehensive. Any gap can be filled by the airline's internal audit function. The complete systems review for the airline can be achieved by the company having its own internal audit section or department or by employing a firm of experts to do the work. Usually the decision on which option to use is related to the airline's scale of operation; larger airlines tend to prefer to have their own Internal Audit Department. Whether the internal audit work is performed by the airline's own staff or by consultants, the scope of the work, its priority and the reports and recommendations produced will all be handled by the board's Audit Committee. Hence, all directors will know about what is going on through the circulation of the committee's minutes.

The internal audit function can help companies and airlines of all sizes; however, many smaller operations are reluctant to commit the board's time to this subject or to the cost involved in having an in-house department. In these circumstances using consultants may be the answer. The cost of using consultants need not be expensive. In a small operation the work of reviewing the systems and recommending improvements should be completed quickly and if it is not because there is some confusion in the system, then the internal audit work will be even more worthwhile. The internal audit review may not need to be done every year as it really depends on what problems are found and the recommendations made after the first review.

The internal audit function, whether in-house or out of house, is just as professional as those of the external auditors, but as it is employed directly by the company it is not as independent. Nevertheless, it provides an extra set of professional eyes and ears. There are a number of professional Internal Audit Institutes in the world which set standards for the profession and issue a Code of Conduct covering the approach and work practices of an internal auditor and they also conduct professional examinations. The purpose of the internal audit function is to independently review and assess the systems and procedures used in the company's operations and to recommend ways to improve their efficiency, effectiveness and security. The focus is on the accuracy of all recording, the application of internal checks and balances (also called 'internal controls') and the identification of risks. All of these will assist the company to reach its strategic goals. Internal audit uses a systematic approach for evaluating the company's risk management, systems of control and corporate governance. These three elements are all interrelated; risk management is part of governance, internal controls are part of risk management and supports governance, and governance provides the framework for risk management and the requirements for internal controls.

The Codes of Conduct issued by the internal auditor's professional institute give guidance on how internal auditors can work within a company and yet keep the appropriate degree of independence in their work. In the real world, the internal auditor and the company both benefit from the internal auditor having an independent but constructive attitude.

'Systems of control' are sometimes referred to as 'internal controls' and are procedures designed to ensure compliance with the company's agreed Code of Ethics and its governance policies and practices. Examples of internal controls are:

- Separation of duties (e.g., invoices for goods or services will be approved by someone other than the person placing the order and payment will be arranged by a third person). Hence, there are three people involved in the one transaction and this separation is seen as reducing the changes of fraud and improving accuracy.
- Establishing levels of authority (i.e., maximum monetary limits for various managers on the value of goods or services they may order or approve). This is to ensure that large commitments are reviewed by the correct level of management.
- Retaining copies of original documents in a central location so they can be referred to if there is a query or investigation.
- Having secure storage facilities for items held in stock as well as keeping detailed records of the number of items received and issued together with a procedure for comparing the physical number in stock to the records, which is to deter theft or holding excessive levels of stock.

In addition to these operating controls there are accounting and financial controls, for example:

- Ensuring the staff responsible for recording revenue are not involved in paying amounts collected into a bank account and that a separate person is responsible for a reconciliation of the airline's bank account records with the statements issued by the airline's bankers.
- Establishing with each of the airline's banks a list of those individuals who are authorised to arrange transactions and those who may approve transactions together with the maximum limit of the transactions they can make each day or week.

- Ensuring that the staff who were not involved in the original transaction independently confirm directly with a bank or a debtor the details of any unsettled transactions.

All of these sorts of controls are part of the airline's governance and are designed to minimise errors, safeguard assets, minimise the risks to the company, maintain ethical standards and reduce the chances for fraud.

The company's internal controls tend to operate at the most detailed level in the organisation, but internal audit will also review the operation of governance in the higher levels of the company. The reviews will include the information, routine and ad hoc, given to the Board of Directors, the process for developing, reviewing and amending strategy, and the operation of the board's committees. The approaches that internal audit may take range from a detailed review of the decisions made including the factors considered and a comparison of the actual results with the expected results, to the softer approach of asking directors and managers for their opinions as to the effectiveness of processes and decision-making. The Board of Directors are responsible for establishing an appropriate system of corporate governance, but are not exempt for having their actions and decisions reviewed, evaluated and sometimes, questioned.

Computer systems can help

The review work of the internal audit function can be greatly assisted and its efficiency improved by introducing computer programs which review accounting records and other information against pre-set yardsticks, and produce reports which list all exceptions to accepted practice or lists of items. Much should be reviewed. Naturally the information in the 'exception reports' (i.e., reports that list any events or transactions that are outside the scope of a defined normal range), need to be further reviewed by an internal auditor or manager to assess the seriousness and significance of the exceptions, but the computer program can do a lot of the hard comparison and collation work. This approach can be used for a wide range of financial and non-financial matters, for example:

- Preparing the reconciliation of the airline's bank accounts (i.e., listing the differences between the bank's detailed records and the airline's). The differences should be reviewed to ensure they are acceptable.
- Comparing physical levels of individual aircraft spares together with their value to monitor whether the agreed minimum and maximum stock levels are being maintained and re-order levels are being used. The comparison can also highlight whether any of the items are now 'slow-moving' (i.e., being issued in smaller quantities or at a slower rate than previously). Hence, these can be potentially unnecessary and can be considered for sale or scrapping.

- Summarising the upgrades of passengers to a higher class on flights, to monitor the effectiveness of the airline's management of passengers as well as to indicate possible abuse by the airline's managers. A similar program can be used to monitor the charging for excess baggage.

It is highly likely that computer systems and artificial intelligence will be used increasingly to do much of the routine and detailed work of internal audit in the future.

Whistle-blowers

In some organisations the Board of Directors involves the company's internal audit function in their 'whistle-blower' procedures because internal audit is required to have an independent view of the company's organisation and its operations. A 'whistle-blower' is an individual (or a group of individuals) who reports an actual or an apparent wrongdoing or breeches of the company's Code of Conduct within a company to the company's most senior management. The wrongdoing can be anything from staff being bullied by their seniors or colleagues, to theft of the company's property. The Board of Directors should establish an effective procedure which enables individuals or groups of individuals to report their concerns about possible wrongdoing, in sufficient detail that it can be effectively investigated, while remaining anonymous if that is what they desire. The report may involve the company's ethics or some other aspect of governance, but in any event the directors need to ensure that each report is completely investigated and that, if there is a problem, action is taken to deal with any individuals implicated and to improve any of the systems involved to try and ensure the problem cannot re-occur. Where there is a situation which may involve criminal behaviour, one of the important decisions for the board will be the extent to which the police authorities are involved. There are advantages and disadvantages to be considered. It is not a requirement in every country that criminal behaviour has to be reported to the police authorities. A proportion of the whistle-blower reports received are likely to result from a perceived problem or a misunderstanding or an unintended sleight, but directors will want to ensure that even these types of reports are dealt with effectively, either to dismiss the report or to change the perception of the action.

The policy for whistle-blowers will usually form part of the company's Code of Conduct and so will be approved by the Board of Directors. As with all other policies, it is not sufficient for policies to be reduced to writing and issued to staff. There needs to be some communication and explanation to all staff by senior managers and directors, detailing the reasoning behind the policy and how it should work in practice. Once the policy has been established and proven to be effective, consideration should be given to extending the policy especially if the company sub-contracts a significant amount of work to other companies.

Business reviews

The area of preparing business reviews illustrates why the Board of Directors need not only to be open to making the changes required by laws, regulations or best practice guidelines, but to be actively questioning whether any of the changes currently being discussed by governments and/or professional institutes and/or by organisations in the industry should be introduced. Indeed, the board should always be looking for ways to improve the company's reporting to stakeholders.

In most jurisdictions there has for a long time been recommendations that Boards of Directors should discuss important general business matters in the company's annual report in addition to the financial figures in the statutory reports. The discussion of the business report is often referred to as a 'business review'. This is not inappropriate because the company's financial result and current financial position are the result of general business decisions. In addition, any commentary which focuses only on the financial figures may exclude a key matter which may have a significant effect on the financial results in future years. Some commentators have argued that a significant number of boards have not followed the spirit of the recommendation or have not done the job well. These comments have generated the pressure to make it a legal requirement in some countries for the board of every company, other than some small and medium-sized enterprises (SMEs), to produce a business review for inclusion in the company's annual report to shareholders. In some jurisdictions the business review is called an 'Operating and Financial Review' or 'OFR'. An 'Operating and Financial Review' is a discussion, prepared by the Board of Directors, of the figures in the company's financial statements in the context of the company's business operations, together with an assessment of the company's prospects, and is included in every statutory report to shareholders. The OFR needs to explain the significant financial figures and relate them to what has happened in the business, then to expand the discussion to include the future of the business, its opportunities and risks. All of this to be done in a way that the average reader can understand the position. Initially this may not be an easy task for the board, but it is no more than the shareholders deserve and have been requesting.

Business model

Any discussion of a business usually includes some description of the company's business model or how the company adds value to its goods and services to achieve a return. There are many definitions of 'business model', some clearer than others. When the Board of Directors goes through the process of describing the company's business model for the first time, they can get some help from the frameworks and guidelines which are available from many different sources (e.g., Institutes of Management, professional bodies and management consultants). Some boards see their business model as an econometric model (i.e., a series of equations which are able to approximately forecast the company's result) while others believe it is a series of general statements defining what it will produce (goods and/or services), how it differs from its competitors and how a financial return is expected to be achieved from these. For the purposes of this discussion an exact definition is not needed.

There are advantages to having a business model which has been written down. Two of the advantages are:

- Getting to the point where the business model is written down means that all directors have the same understanding and this is important whether the model is for an operation which is just starting up or for one that has been functioning for many years.
- The model can be reviewed, discussed and possibly changed.

The second of these is central to the business and may be existential because the model deals with the future and directors are concerned with managing and guiding a company's future. History and certainly recent decades have shown that periodically business models need to be modified or completely changed. For example, the road haulier with a horse and cart, when looking at the recently developed steam locomotive, should see a reason to considering whether their business will change; similarly taxi operators when they see ride sharing; operators of full-service airlines when competing with no frill carriers. These are just a few examples of where a current business model may need to be re-considered by the Boards of Directors and, possibly modified, so the company has a viable future. It is possible to change a business model and almost always the initiative needs to come from the Board of Directors. In a joint-venture an individual shareholder or group of shareholders may also suggest a review, but generally the board takes the lead. Directors need to be aware of the changes which are happening in their industry and also in other industries, to try and get a sense of what effect the changes or proposed changes might have on the company's operations, hence its results, and then to consider how the company might react. This is a part of the board's essential duties and its responsibility for governance, strategy and risk.

If ever or whenever changes are made to the company's business model the effects of the changes will probably need to be reflected in the company's long-term strategy and the supporting strategies (e.g., Finance, Customer Service). It may also be necessary to make other changes, for example, to the terms of reference of some of the board committees to take account of the new approach. The change may be particularly important to the Risk Committee; a changed approach to the business probably means changes to the company's business risks. Once the full effects of a change to the business model have been thought through, the change and probable effects need to be communicated by the directors to all staff. The board should anticipate questions from the staff because any changes are almost certain to mean that the way staff work will have to change and in addition the way their performance is assessed may also change.

These difficulties can be anticipated and should be dealt with as sympathetically as is possible in the circumstances.

There are many ways that directors, individually and collectively, can stay in touch with business changes. The individual directors, who are INEDs, bring with them knowledge of other industries and the possible and likely business changes. Part of each director's Continuing Professional Development will include reading industry magazines which will, without doubt, discuss the changes being considered or discussed within the industry or which may influence the industry. Directors should be encouraged to attend seminars on developments in the industry as well as on general business matters, for example, on 'artificial intelligence'. This is another area with many definitions, but a useful one is, 'the development of computer-based systems which do the jobs normally done by humans'. On some subjects it may be worthwhile asking an expert in a particular field to discuss a subject with the directors, and possibly senior managers, at a board meeting. Whichever of the options is taken, the responsibility of ensuring the company's approach to its business remains relevant, rests with its Board of Directors and is part of their governance duties. It is unlikely that the board alone will have the resources to examine in detail the possible financial and commercial effects of a change in the company's business model and it may be sensible for the board to hand the project over to the director who oversees the company's planning process and ask for a report and recommendation which can be subsequently discussed by the board.

Comparisons with others

The Board of Directors, in addition to keeping an eye on what is happening generally in the industry to see whether there are processes and approaches which would benefit the company, also needs to have a sense of what other companies are doing, both in the same industry and in other industries.

To some extent the need to look at other industries is covered by keeping the business model under review, but that will not cover all cases. Other industries may provide the same or similar service more effectively or efficiently than the airline industry or some airlines in the industry. Other industries may be dealing with the same problems as found in airlines, but in a different way. The board's responsibility to monitor performance almost by definition involves them in comparing the airline's financial and operating performance with others. The airline may be performing well when judged against its own plans, but not well when compared to what other airlines are achieving. The long-term trend is for the real cost of travel to fall for the traveller, and this imposes on all airlines the need for continuous improvement in efficiency and effectiveness and a focus on what the customer values.

Any comparison should focus on the KPIs which the airline has. The KPIs are related to the essence of the business and this is the area that needs to be right, or more correctly, at its best. Airlines need to benchmark themselves against their competitors and other service-providers. For the sake of easy reading, 'benchmarking' is the process of making comparisons on service and methods with other companies which are considered to be successful, whether in or outside the aviation industry. This definition may not be completely correct, but it is near enough to be useful. Benchmarking can be done:

- Within an airline, for example, comparing the on-time departure records of each port in the network.
- With other airlines, for example, comparing each port's on-time departure record with the published figures for other airlines.
- With other industries, for example, with the on-time departure performance of trains, buses and cruise ships.

The objective is not to be self-satisfied, but to identify those who can and do perform the same or similar operation most effectively and efficiently.

The idea of benchmarking may initially seem strange or difficult. Part of the planning for any new airline will usually involve assessing how 'good' airlines perform their operations partly to help the start-up one assess the resources it needs, but also as a model for the new airline. Benchmarking in an established airline can be thought of as updating that initial model and approach.

Benchmarking does not usually produce a perfect comparison, just a very good indication, but one that can be useful. Comparison of both the financial performance and operations of other businesses relies on information and that is not always easily available. Financial information is in the airline's or other company's annual report, but that will only be freely available if the company distributes its annual report and there may be differences in accounting polices (e.g., depreciation), that need to be adjusted for, and the information may need to be summarised to be useful. Some operating information is published generally (e.g., on-time performance), but even here there may be differences in how the figures are compiled. Some airlines use the industry standard of any departure being 'on time' if the aircraft leaves the stand within 15 minutes of the scheduled departure time, while others compare with the scheduled departure time. Still others exclude any late departures due to severe weather or industrial action, so at best the information is an indication, but it is nevertheless an indication that can be useful. Trade associations like the International Air Transport Association (IATA) have information on many aspects of the airline business and although it may be aggregated so that individual airlines cannot be identified, useful indications of average performance can be found. An examination of the available information on the recipients of industry awards can also yield some useful guidance. All-in-all it is possible to get some view into the way the 'best' airlines operate and it is important that directors are not discouraged by the initial difficulties in getting information or good indications.

Making comparisons with other industries can also be difficult. Many companies will be open to an approach to exchange operating information because the companies are not in competition, but the difficulty is in deciding what in each company can be beneficially compared. There may be companies in other industries which are better at queue management (useful at an airport) or communication with large numbers of people (passengers waiting during a weather delay) or advising customers of deliveries (air-cargo). If the airline can learn something beneficial it should do so. The detail of this is not a matter for the Board of Directors directly, but they have an interest in ensuring that the work is done, as it is part of monitoring the airline's operations and efficiency as an implied requirement.

The Board of Directors is required to keep its eyes on the operations of the company. This means looking at past results to learn lessons, monitoring current operations compared to plans to ensure everything is on schedule, describing the desired future and organising the resources to achieve it. Fortunately, the Board of Directors can establish systems and procedures which will help them deal with these matters, and each director needs to take an active role. The systems and procedures should not be allowed to become a box ticking exercise, and governance must remain an energetic and open process; all aspects of governance need to be alive and thriving.

9   What comes next?

Leaving

At some point every director leaves the board. The Board of Directors' Nomination Committee, working with the board Chair, will be as active when a director leaves the board as they are before a new director arrives. In the vast majority of cases directors leave the board 'on time', either when they are due to retire or they have said they wish to step down. It is an exception for a director to resign from the board unexpectedly, but the Nomination Committee must consider the possibility of the unexpected as part of the company's risk management programme. The reasons for an unexpected resignation can range from a new significant conflict of interest to illness.

It is even more unusual for a director to be removed from the Board of Directors by the shareholders or for a legal reason. The shareholders appoint all directors to the company's board at the company's AGM. In practice they generally endorse the recommendations made by the Nomination Committee supported by the current Board of Directors. At the same meeting the shareholders will also approve the re-appointment of a director who has reached the end of their initial period and are to serve for a further term. If for any reason any of the shareholders do not wish an individual to be a director, they may vote against the appointment or re-confirmation. The decision will be made by a vote of all the shareholders. Rejection of a nomination for a directorship or not re-confirming an existing director is very unusual, because usually the Chair and the board will have been made aware earlier of any shareholders' dis-satisfaction and have taken action to avoid the negative vote. A majority of shareholders can remove a director at any time by giving notice to the company and calling a general meeting of shareholders to pass a resolution removing the director. The period of advanced notice that the shareholders must give before a meeting of shareholders can be called and before the meeting, varies between jurisdiction. The director involved may speak at the meeting to rebut any claim against them or to explain their position. The director may also give a written submission, but is not required to do so. A side issue of the process of removing a director is that if this happens it points to a serious weakness in the company's governance, either in the communications with shareholders or the process of regular board assessment of the Chair and the board, or both of them. The Chair and the board will need to address these issues as soon as possible.

There are other circumstances when a director must leave the board. These are mostly legal; some being contained in the law and others in the company's Articles of Association. Typically, these are if a director is:

- Being judged bankrupt.
- Ruled as being mentally ill.
- Convicted of fraud or other criminal offence.
- Persistently failing to submit required government forms or returns.

These may vary country by country and additional reasons may exist in other jurisdictions.

There are two main reasons why a director may leave the board earlier than planned: due to ill-health or a disagreement within the board. Resignation for health reasons just has to be dealt with whenever it occurs, but a dispute on the board needs some care. Differences of opinion can arise at any time during discussions in the board room. Directors are required to act independently and to test proposals and plans submitted to the board for approval, but in normal circumstances the differences can be resolved and settled so that all directors can support the final decision. Even if a director is not completely won over by the opinions of the other directors, once the board has reached a decision, all directors must support it, however difficult this may feel. It does happen, but it is very seldom, that a director feels so strongly on an issue that they cannot accept or support the board's final position and decision. In these circumstances the only course is for the dissenting director to resign from the board. Once the director makes the decision to leave the board, the communication outside the board needs to be handled carefully, especially if the company is publicly quoted and an announcement has to be published. The responsibility for the communication normally falls to the Chair, often supported by the company's Communications Department or a consultant.

Whatever the circumstances of a director leaving a board are, it is not usual for the ex-director to receive any termination compensation. If there is a proposal to make a termination payment, the proposal should be put before the shareholders for their decision at a general meeting.

When this type of unplanned departure from the board happens, the company's Articles of Association usually permit the Board of Directors to appoint a new director to fill the 'casual vacancy', which is an unexpected vacancy, for the remainder of the period the original director was due to serve. Shareholders will be asked to confirm the appointment of the new director at the next AGM. If the director who resigned has appointed an alternate director, it is not unusual for the alternate director to be appointed as the replacement, but this depends on the skills and experience currently needed on the board and whether the alternate director wishes to accept the appointment.

Regardless of the reason behind a director's departure from the Board of Directors, it is necessary to complete forms and to report the change to government departments. This will generally be handled by the Company Secretary. It is important that the formalities are completed promptly because there will be fines for late compliance and the fines may apply to each day the returns are overdue.

There is one last area for the departing director to deal with and that is what should be done with the confidential company papers the director has received either physically or in electronic form, during their time on the board. For safety's sake it has to be considered that all information received is confidential, hence the ex-director has a continuing duty to safeguard it. The safest course is for all the information on paper to be returned to the Company Secretary for destruction and for the electronic information to be deleted from the retiring director's computer records. The company's preferred course will usually be decided by the Company Secretary.

The future

The scope of corporate governance has developed significantly since the first company was formed. The changes tend to be made slowly and are often in response to some form of substantial business problem like a major bankruptcy. The spark for change is frequently a comment or recommendation in the conclusions of an official report into the causes of the business problem. Frequently a report into a major business failure will indicate that the problem could have been avoided or the effects mitigated, if the Board of Directors had taken certain actions, had monitored a specific area of the business more closely or had a certain reporting system in place. The reaction to this type of comment is often to produce additional recommendations on the best corporate governance practice and to further clarify the responsibilities of company directors. So, something good and useful comes out of the analysis of each business failure or problem; a cynic might say that changes in corporate governance are only ever made in response to corporate disasters.

It is helpful to look at the problems which are being experienced, in varying degrees, in the area of corporate governance now. All is not sweetness and light, and there are problem areas. Companies are important to the world's economy and solutions to their problem areas will be found. Reviewing and discussing the current worries is both very useful for anyone considering becoming a director and who wants to understand the problems of how corporate governance works now. It helps to give a more balanced view of directorship and better prepares a new director for their role. This final section reviews some of the current anxieties there are amongst directors and potential directors. These are all open issues, so sadly it is not possible to suggest a solution for each problem. Some of the issues are considered to be sufficiently serious by some individuals that they have declined to accept a position as a director on a board because of their concerns for the problems. Although directors are included in the company's liability insurance cover, it is not possible to insure against the loss of an individual's reputation and it is possible to do everything by the rules and still suffer some damage to one's reputation in the eyes of the public. Public opinion is not the same as a court of law.

Responsibilities

Perhaps the most difficult question to answer on corporate governance is 'What are the responsibilities of the Board of Directors and of individual directors?' Depending on the source, answers can be a:

- Long, 20 plus, open-ended list of 'the main responsibilities'.
- Shorter, but still open-ended, 'summary of the key responsibilities'.
- General over-arching phrase like '…is responsible for managing the company and achieving its success'.
- General image of the role (e.g., 'the board is like the captain of a ship responsible for directing the ship and for the safe and successful completion of the voyage').

None of these fully answers the question. The decision to become a director requires acceptance of the uncertainty of exactly what the full list of a director's responsibilities are.

This can lead to the further question of 'If the responsibilities are open-ended how will I be judged, if something goes wrong?' If the 'something' ends with some form of legal case against a director for neglect it is likely that the concept of a 'reasonable man' will be used to judge whether the director or directors acted appropriately. The 'reasonable person' (formerly 'reasonable man') concept is useful when it is not possible to specify exactly what should have been done in every circumstance. It poses the question, 'What would a reasonable person, exercising average skill, experience, care and judgement, have done in these circumstances?' This sets a standard for judging the actions actually taken. People with specialist skills or experience are usually judged slightly differently because of their special skills. Even this concept is not a great deal of comfort to a potential director because it is also not a precise definition either.

Every board should define and record its relationship with the company's operating management. This is usually done by the Board of Directors formally delegating specific responsibilities to the 'Managing Director' or 'Chief Operating Officer', including limits on the amount they can spend on capital items etc. In a bicameral set-up the delegation is from the Supervisory Board to the Management Board. In addition, the board should agree a list of decisions which can only be taken by the Board of Directors. These are often called 'reserved powers' and can include such matters as buying and selling assets, and agreeing borrowings. These documents may help the CEO and operating management, but the Board of Directors is still left with the remaining open-ended responsibilities.

A board's responsibilities are often further defined after some form of commercial disaster has been investigated and the cause discussed. There is no reason why this process should change in the foreseeable future so the Board of Directors and each individual director must be prepared for changes and for additional responsibilities to be clarified and added to their list periodically.

Communication with shareholders

In recent memory there have been some disputes between a company's shareholders and its Board of Directors. The majority of the disagreements that have been published have been in publicly quoted companies probably because their AGMs tend to be more open. This focus on public companies does not mean that the shareholders of private companies do not have the same concerns; it is just that the exchanges of views at the AGMs of private companies have not received the same publicity or that the board has been able to resolve any issues. The pressure for a more open and frequent communication between the board and shareholders comes not only from shareholders and shareholders' groups, but also from organisations like government departments interested in regulating companies' financial operations, trade unions and special interest groups, so the pressure is from the broader group of stakeholders as well as from shareholders.

It is easier for the Board of Directors of a private company to communicate with its shareholders than for the board of a public company because the stock exchange's rules will require that a public company make public any information which may influence the price of the company's shares. Most of the issues shareholders have are very likely to be on 'price sensitive' matters (e.g., strategy or level of dividends). In addition, there are likely to be significantly fewer shareholders in a private company than in a public company and there are also probably significant differences between the make-up of the shareholders in a public and a private company. Public companies will be a mixture of individuals, institutional investors (an organisation that invests in other companies either on behalf of others or for its own account) and investment managers; the objectives of each investor may also be different with some interested in the company's potential to produce a regular income, others looking for an increase in the value of its shares, yet others looking for both. The shareholders in a private company are more likely to be homogeneous. Also, it is often easier to talk to a small group than a large group.

The main issue which shareholders have raised and which has been publicised has been related to directors', usually executive directors', and senior management's remuneration (i.e., their total reward package including salary and all other benefits including pensions). A second publicised issue is usually related to what the company's strategy is and should be; this usually comes to the fore when a company's financial performance does not meet shareholders' expectations. Another matter of concern is the company's effect on the environment. Naturally other issues arise as circumstances change. There have been instances where the board have simply refused to discuss these issues with shareholders and this has caused a major disagreement.

The well-publicised disputes on remuneration have, in some jurisdictions, led to regulations being introduced requiring a company's Board of Directors or its Remuneration Committee to give shareholders a detailed report on directors' and executives' remunerations. Often the report is 30 or more pages long and covers the current policy with any proposed changes, details of short-term and long-term incentives for individuals together with details of when the incentives are payable, details of any targets which can influence an individual's remuneration and, in some circumstances, the details of individuals' remuneration. Acceptance of the report may have to be voted on by the shareholders.

In some jurisdictions there are regulations applied to public companies imposing further requirements on the Board of Directors if the Remuneration Committee's report is rejected by a significant minority of the shareholders; frequently 25% or more. If a significant minority reject the Remuneration Committee's report in more than one year (e.g., two or three years), there may be a requirement for the entire board to resign.

Preparing the remuneration report can be an onerous task and the resignation of the Board of Directors is a significant event. An effective board and one which understands its fiduciary duty will anticipate shareholder issues before they arise and have a plan for dealing with them. It is better to be pro-active than reactive.

In normal circumstances, a Board of Directors only meets its shareholders once a year and then mainly to discuss the financial results for the previous year and to complete the legal formalities. In these circumstances, if there is not the chance for the shareholders to openly discuss the business and any concerns they may have, it is easy for a feeling of distrust to develop over time and this cannot be constructive. It needs to be remembered that the directors are responsible for running the company on behalf of the shareholders, so both groups should, ideally, be of one mind.

There is little, if any, guidance on the best or preferred way for the Board of Directors to communicate with shareholders. Some companies, possibility a minority, both public and private, have tried different approaches, but no one approach is yet judged as being completely successful. Ideas that have been tried include:

- The company having a facility on its website where shareholders can question the board and receive explanations. Alternatively, they may appoint a named individual with published contact details to fulfil the same function. While this may satisfy an individual shareholder, it means that only the shareholder who asks the question is given the answer, even if other shareholders have the same query.
- A similar approach is to have an informal 'meet the directors' session after the AGM, and while this helps shareholders build a relationship with the board, it has the same disadvantage that it is unlikely that everyone's queries

will be addressed and any constructive suggestions from a shareholder are only known by the director who hears them.
- Hold meetings with shareholders other than the AGM, at which all the directors listen to shareholders' concerns and observations. It is useful if after this type of meeting a summary of the points raised is sent to all shareholders, so there is a record and those shareholders who did not attend the meeting are kept up to speed.

If a company has a large number of shareholders, there is the potential difficulty of finding the right location and format for the more informal meetings between the board and shareholders. Some companies with a larger number of shareholders have tried meeting a smaller group of shareholders restricted to those holding a minimum of a certain number of shares. These may be called 'key shareholders'. The drawback of this approach is that it places the smaller shareholders at a disadvantage and discriminates against them.

Another option for improving communication with shareholders is to appoint an Investor Relations Manager or to establish a department to deal with investor relations. Having a specialist manager or department can be a very useful source of advice and information on techniques and approaches, and also on ways to create a plan for developing open communication with shareholders. However, there is the drawback that it is extremely difficult to build up shareholders' confidence in the Board of Directors by using a messenger. The directors have to interact with shareholders directly in some way and the efforts of directors to improve communications needs to be seen as a sincere personal endeavour.

There are some organisations which publish on their website an abridged copy of the minutes of the meetings of the Board of Directors as a way of including shareholders and stakeholders in the communication loop. Discussions on commercially sensitive matters will be omitted; the current in-phrase is 'redacted' (i.e., to have removed or blanked out parts of a report). This approach is popular with government organisations (e.g., Trade Development Councils) and not-for-profit organisations, but there are no practical reasons why commercial organisations should not consider using some form of this approach.

Any communication plan should be part of a more general communication strategy by the Board of Directors with staff and stakeholders. The plan needs to deal with routine communications on such matters as annual results, but also ad hoc issues like major accidents. The plan should also have an agreed approach on the line between confidential and non-confidential matters. The main thrust of the plan will be to increase transparency of how the board makes its decisions in order to increase confidence in the board's processes. It is probably best for the board to gradually increase its interaction with shareholders, trying to establish which approaches work well and which do not, but there are some who advocate launching a fully blown communication policy and plan at one time in order to impress on shareholders that 'things have changed for the better'.

134 *What comes next?*

Friction with shareholders and stakeholders should be minimised and, if possible, avoided all together because everyone has a vested interest in a company's success, however defined. Open discussions between the board and shareholders can certainly help but the trick is to find out what works best for the company and for the shareholders.

Reporting

One of the Board of Directors' duties, both legally and socially, is to report regularly to shareholders on the company's results and its current position. Best governance practice extends the reporting to include other stakeholders as well.

The content, and to some extent the format, of company reports, is now driven by international reporting standards and, where the company's shares are quoted on a stock exchange, by the stock exchange's reporting rules. Many readers of company reports have found that in recent years the reports have grown in size and complexity and are becoming more difficult to understand. There are some company annual reports which are more than 300 pages long and weigh more than a kilo, and many have 150 to 200 pages. In an effort to try to meet the highest standards and still make their reports easy to read, some companies have split their annual report into separate volumes, each dealing with a separate subject such as financials, environment, governance, etc. The information and format are the same as in a one-volume report, but it is easier for the reader to find the information they are interested in. Where there are separate volumes it is not unusual for the company also to issue a brief summary of the key information from each volume, rather like an executive summary, and often this is called something like 'At a glance' or 'In brief'.

Other companies have taken the approach, where this is legally permitted, of including on the company's website the information which does not change frequently (e.g., a description of how the Audit Committee is constituted and works), and including in the company's annual report a comment directing the reader to that area of the company's website. The sort of information which is suitable for this approach is the description of the governance structure, the board, board committees, policies, etc.

Generally, it is easier for a large company to arrange to have the resources needed to report to the highest standards. It is necessary not only to understand the reporting standards but also for the company to have its own systems which produce the information needed for the reports. Smaller companies may find complying with the standards much more difficult and in many cases, there are special rules and/or exemptions for companies below a certain size. However, the result can be that the reports of smaller companies do not contain all the information that some stakeholders would like.

The bodies charged with devising and advising on reporting and communication with shareholders and other stakeholders, such as the International Integrated Reporting Council, International Financial Reporting Standards Board, Global Reporting Initiatives and Sustainability Accounting Standards

Boards, continue to co-operate and work on refining their advice and standards in the light of the experiences of companies and to accommodate the desires of the readers of company reports. In addition, there are other separate organisations which are 're-thinking' the whole subject of reporting by companies and their communications with all stakeholders.

It might be tempting for the Board of Directors to do the minimum required by the law and regulations and to leave 'reporting' at that, but there can be advantages to reporting to a higher standard than simply the minimum and having the systems which can produce information required rather than the minimum. The advantages of better reporting can include greater confidence in the board and company by staff, suppliers, banks and other financing institutions, which may lead to better terms. Having enhanced underlying reporting systems may mean that the board has additional information which will help them when they make their decisions and with the information which can help minimise operating costs.

This situation can pose problems for the Board of Directors, its Audit Committee and possibly its Finance Committee. Directors need to stay up to date with the advice and standards being issued and to be aware of the need to have the internal systems required to produce the information needed.

Ways to achieve the desired balance between information and readability have yet to be found universally and it is likely that 'reporting' will be an issue for the board for many years. Reporting is the responsibility of the board; the annual report is their report to shareholders.

Using technology

The advantageous use of developments in computer systems and their application to the business is becoming an issue for Boards of Directors. The amount of intellectual energy needed on the subject of technology varies between industries, but airlines have been and are likely to be significant users although all of the economy will be affected to some extent. There are new technologies being developed and implemented in business and their full potential and effects cannot yet be foreseen. There are many established and developing technologies, some of which may help businesses. Those discussed most frequently now include:

- 'Artificial intelligence'.
- 'Big Data' (i.e., the analysis of large sets of data to establish trends and preferences).
- 'Internet of Things' ('IoT') which is a network of a variety of machines able to transfer data without human intervention.

The Board of Directors has to stay in touch with these major developments and any others that may benefit or influence their company. The board does not need to understand the fine detail of each type of technology or exactly how it functions, but they do need to understand the possible benefits to, and effects on, the company's business as well as the possible change in the company's business risks.

Assessing the potential of any technology can be very time-consuming, especially if the technology is very new and there is little operating experience. The Board of Directors has to find a way to be kept up to date with developments while using their time effectively. Whether this means appointing an individual with specialist knowledge of technical developments as a director, either executive or non-executive director, or establishing a group to monitor technology developments, is a decision for the board. There are disadvantages to expanding the size of the board to include experts who may not be experienced in business and another option is to appoint an advisor on technology to the board, but this means that the majority of the information will come from just one person. Forming an advisory group to the board may appeal and could work as could forming a board committee or a committee of senior managers from within the organisation to monitor developments. If a group of some type is formed, care needs to be taken that all of the members' opinions are reflected in the reports to the board not just one opinion or recommendation, because many of the reports will be commenting on possible future uses, risks and problems, each of which will have different probabilities. The more options that are communicated, the more important it will be for individual directors to be able to evaluate the arguments and this may require some amount of training.

Whichever course is taken, it will be important to include the responsibility for assessing the potential risks of implementing and not implementing the new technology in some form in the Risk Committee's mandate.

For an airline, technology is a particularly important area because by the nature of its operation an airline handles large volumes of data on passengers, cargoes, spares and supplies. In addition, there are many dealings with individuals, passengers, shippers, the general public as well as suppliers. Not all of the dealings will be in calm and relaxed circumstances and the airline's reputation is formed by how each situation is handled. If in difficult circumstances some form of technology can increase the chances of a good result, that will be attractive, but if the technology only benefits the airline and not the individual, that may not be such a good outcome.

Operating safety is paramount in an airline, whether the airline is large or small, and nothing should be done which reduces the airline's operating safety standards. This means that the airline's whole safety structure needs to be involved in the assessment of any new technology. In the context of new technology, 'safety' means not just flight safety and operating safety, but also areas such as the safety of passengers' data and the airline's data. Fortunately, the opinions and themes on 'safety' can all be drawn together in the airline's Risk Committee, or whichever board committee is mandated to deal with safety.

Decisions as to whether to agree a proposal to introduce a new system or technology are not always straightforward. The easiest decisions are those where there are clear monetary benefits, and in these cases often the money makes the decision and other benefits are considered to be incremental benefits. Not all proposals can show clear monetary benefits which can be identified, isolated and reported, such as for example, advanced weather radar or a more comprehensive rostering system. In these cases, some other measure needs to be applied to the expected outcome of the proposal, such as whether the new technology can be shown clearly to assist in meeting some strategic goals which will be reflected in one or more of the airline's KPIs. The type of proposal which does not have identifiable monetary benefits is frequently encountered in an airline and there should be an established procedure for handling the proposal, because if there is not, then a procedure needs to be devised.

Using new technology has the potential to benefit many businesses in all the areas of their operations, not just financially by increasing revenue and/ or reducing costs but also in safety, customer satisfaction, staff relations and the companies' effects on the environment. Monitoring and evaluating new technology often presents significant problems for the Board of Directors and there is no established way to handle the subject. However, this does not mean the board can ignore it. A way that makes sense to the board and to the business has to be found. There are guidelines for applying governance to the technology function, but it is difficult to get advice on the governance techniques for applying technology to a business.

Relationship with employees

In every organisation the staff are a very important element in achieving success. In companies like airlines, a large number of the staff are actually part of the airline's product and are in constant contact with customers and can influence their attitude to the airline. Although not directly involved in the day-to-day management of staff, the Board of Directors has a significant influence on the relationship with all employees.

The atmosphere and culture of the company is directly and indirectly set by the Board of Directors, partly in the policies and practices approved by the board and partly in the way the board as a whole and as individual directors act. The wording and content of the company's policies and approved practices set guidance and limits for the behaviour of staff and management as well as its directors. Effective and open communications by the directors with staff means the staff can more easily become involved in the company's strategies, policies and business ethics. If staff are involved with the company, they are more likely to make a greater contribution to achieving its strategies and maintaining its good reputation.

The boards of smaller companies will probably find it easier to keep a close relationship with their staff than larger companies simply because of the different scale of operations. In a typical airline, more than half or about 60% of its staff are flying crew, with the actual percentage depending on how much work such as ground handling is sub-contracted to third parties. In addition, it is likely that there will be a significant number of staff working away from the airline's head office, both nationally and internationally. Ensuring good communication with staff in (say) an international airline with 10,000 staff, when many of them are away from the home base for most of the time, is more difficult than in a small domestic airline where most of the staff come back to base every evening and, possibly, where the directors know some of the staff by their first name. Despite the problems, the larger the airline the more important effective staff communications become. Regardless of an airline's size, good staff communications are essential for success. The board needs to remember that 'communication' is not just issuing circulars or writing articles in the company's staff magazine, but meeting and listening to staff.

The executive directors on the board have a particularly important role to play in staff relations. How the staff act and make decisions in the everyday management of the business should follow and therefore endorse, the business policies and practices approved by the Board of Directors. The executive directors are in day-to-day contact with staff, and hence have the chance to explain and answer questions on the airline's strategies, policies and desired practices. These all help to build a good working relationship with staff and trust in the board. Being open to speaking to staff and listening to their views and concerns can not only lead to a relaxed working atmosphere, but importantly can clarify any misunderstandings and can correct any rumours which are circulating.

The non-executive directors also have a role to play in the relationship with staff. Non-executive directors generally have a more distanced relationship with staff because they have less interaction with them day-by-day. This may mean that non-executive directors' relationships with staff will be less relaxed than those of the executive directors, but in a curious way may also mean that it is very likely that the words and comments made by non-executive directors may carry more weight and authority. A successful business is usually one where the board effectively communicates with staff on key issues. The communication should involve allowing time for staff to give their own feedback and views on topics. A board that is not open to discussion on the company's strategy and results is likely to create an atmosphere where rumours and speculation replace strategy and fact. Non-executive directors should be prepared to speak to groups of staff regarding strategies and results, and listen to their feedback. The feedback from staff to both executive and non-executive directors should be collated and discussed periodically at board meetings.

Good communication will ensure that all staff clearly understand the airline's strategies, policies, results and current priorities. Staff with this information can work towards achieving the airline's goals. The process of communication which includes listening to staff should motivate them and ensure the whole organisation is working towards the same goals.

The company's staff employment conditions and how they are administered are important motivators and the conditions and their administration should be aligned with the airline's goals. The board's Remuneration Committee should make recommendations in both of these areas. How employment conditions are administered is an important element in creating a positive atmosphere in the company.

In resolving the problem of how to achieve a beneficial relationship between the company and staff, goodwill and good intentions alone are not sufficient. The problem of communicating with a largely absentee workforce, whether flying or working away from head office, was discussed in Chapter 4. Sadly, although there are plenty of options, there is no sure-fire way to achieve the objective. Improving a relationship depends partly on what the relationship is today, and each airline needs to find its own way of establishing and maintaining good relations and open communications with all its staff.

Retirement plans

The responsibility for a company's retirement arrangements are not often discussed in the context of corporate governance, but it is an important issue for the Board of Directors because it involves staff morale and motivation, the funding position of the company and the company's reputation.

A pension is generally a payment made regularly to someone previously employed by the company, who is now retired because they exceed the agreed retirement age; a pension may also be paid on early retirement due to ill-health. In some jurisdictions a pension is paid by the national government and is not a matter for companies. In other countries all companies are expected to pay pensions to their staff when they retire, in yet others there is a mixture of the two systems. The annual cost of pensions is tending to increase because of increasing life expectancy and those governments who pay pensions out of general tax revenue are becoming inclined to encourage companies to take over the responsibility, because of the increasing liability and to prevent general tax rates from rising.

The subject of pensions and retirement payments tends to be complicated, because individual schemes tend to be complicated. The detail of the schemes frequently has to be designed to follow local tax laws and agreements negotiated periodically with groups of staff. The complications and agreements surrounding retirement plans together with possible legal requirements which vary between jurisdictions make the subject a difficult one for the Board of Directors. Regardless of the complexity and the difficulty, the Board of Directors remains responsible for monitoring the condition of the company's retirement plan(s). Because of the complexity of the subject, this section can only deal with the basics of retirement plans and highlight the potential problems that the Board of Directors may face.

Essentially there are two ways of calculating the amount to be paid to a retiring staff member. One, a 'defined benefit scheme', bases the pension amount on the final salary and years of service of the staff. The alternative is a 'defined contribution scheme' where the company makes regular contributions to a fund for each member of staff; the staff member may also have the ability to make contributions. The fund is invested and the income re-invested and at retirement the total amount in the staff member's fund is used to buy a pension for that individual. They may also have the option of taking part or all of the fund in a lump sum. There is also a 'provident fund' where a lump sum payment, based on salary and years of service, is made to staff when they retire and the company has no further obligation to the member of staff, as the individual is responsible for making their own pension arrangements.

Whether the company pays a pension, contributes to a fund or pays a lump sum, it is likely that the amount calculated to be the company's liability will be held separately in some form of entity, often a trust, and its investment will be managed by a third party independent of the company. Often there are tax incentives which encourage the company to separate its retirement scheme from the rest of the company's operations. Even if there are not any tax incentives, it is good practice to have an independent fund for retirement because it:

- Gives staff confidence that there are funds available to pay the retirement benefits and reduces the risk of not receiving any retirement benefit if the company gets into financial difficulties.
- Removes the liability from the company's Balance Sheet.
- Means the separate fund can be invested to produce a return which increases the amount of the fund.

All of the approaches which require the company to provide a retirement package have problems for the Board of Directors:

- Pensions payable at regular intervals are an open-ended liability for the company because the amount payable depends on the actual remaining life of the company's retired staff. Each year the company needs to put aside sufficient liquid funds to cover any increase in liability based on a calculation of the estimated life expectancy of all its retired staff and the expected return from investing the fund. If the retired staff live longer than was originally expected and/or the return achieved by investing the funds is less than forecast, the retirement fund will need to be topped up to cover the shortfall.
- A defined benefit scheme has similar drawbacks to a pension, as the company must top up the fund each year to take into account any increases in salary together with the extra year of service. It is likely that the scheme's funds will be invested, but if the value of the investment falls, the company will have to make good the loss.
- In a defined contribution scheme, the amounts transferred to the retirement fund are generally invested and the staff get the benefit of any appreciation in the fund, but carry the risk of any loss. If there is a substantial loss or series of losses in the retirement fund, there may well be pressure from staff for the company to make good, or to contribute to, the loss.
- The calculation of the amount of a company's obligations to a provident fund is perhaps the easiest. However, if the company makes annual

contributions to the provident fund assuming that the accumulated fund will be invested to earn a return to be used to reduce the company's contribution in subsequent years, the company bears the risk of fluctuations in the investment performance of the fund.

In recent years the trend has been for companies to close their provident fund, defined benefits schemes and pension schemes and move staff into defined contribution schemes.

The problems with retirement schemes for the Board of Directors arise mostly from the administration and investment of the separate retirement fund and any possible open-ended liability to top up funds. Although the retirement fund is separate from the company, the beneficiaries (the staff), are still employed by the company and any poor performance in the retirement fund is likely to have an effect on staff morale. For this reason, the company commonly appoints the trustees or managers of the scheme and probably will have a significant influence on the retirement fund's investment managers and advisors in an effort to ensure the highest standard of administration and prudent fund management. It can be difficult to find individuals with the desired experience and knowledge, particularly when the retirement fund is large. Any problem with the fund, not just any large loss, will reflect badly on the company's board even though the fund is separate from the company.

The company's annual board plan should include regular reports, at least once a year, from the trustees or managers of the retirement fund in sufficient detail for the directors to assess the likelihood of a problem developing. This type of review requires directors to understand the investment market and ways to minimise investment risks; this is not easy especially in volatile markets. To assist the board, they may periodically commission a specialist company to review the operation and the performance of the retirement fund together with the level of risk. It is likely the pressure for companies to provide retirement benefits for their staff rather than for governments to provide them, will continue to be felt by companies.

Problems

The objective of mentioning these current corporate governance difficulties is not to make the process seems more difficult than it is, nor to discourage anyone from becoming a director, but to try to give a balanced view of the subject and indicate where the likely focus for changes and developments may be. Except in the cases of fraud or other criminal behaviour, the vast majority of boards do their work diligently and without the threat of litigation. The boards make good decisions and make some mistakes just like any other group of individuals.

Taking a formal approach to corporate governance with policies, committees and reports is focused on adding some structure to the human process of guiding and managing a company towards achieving its success.

Glossary

These are the definitions used for terms in this book. In other circumstances and books a different definition may apply.

Accounting practices   A list of procedures used to record transactions in the accounting and books of account agreed by a company's directors and management and used when preparing the company's statutory and management financial statements. Accounting practices are not the same as accounting standards but are the way the company has interpreted accounting standards or has decided to deal with an accounting matter not covered by an accounting standard.

Ad hoc   In business it usually means 'when needed' or 'when requested'.

Advisory Board or Advisory Panel   A Board of Directors may decide to establish an 'Advisory Board' or 'Advisory Panel' to give advice on specialist subjects upon request. Each advisor will be an expert in a particular subject. The advisors are not directors and cannot vote on any board decisions.

Aircraft utilisation   The average number of hours during each 24-hour period that an aircraft is operating to generate revenue.

Air Operators Certificate (AOC)   The certificate issued to an airline by the local regulatory authority permitting an airline to operate specified aviation activities. To secure a certificate the airline usually has to satisfy the licensing authority that it has the staff, assets and systems to ensure the safety of the general public and its own staff.

Alternate director or Alternative director   An individual appointed by a director to stand in for them should they be unavailable to attend a board meeting. An alternate director is empowered to perform all the duties of the director at the meetings the director cannot attend.

Annual General Meeting (AGM)   A mandatory meeting open to all shareholders at which certain legally required business is conducted. The company's Board of Directors presents a report on the company's result and strategy, and the shareholders exercise their rights (e.g., electing directors).

Annual report   A report to the shareholders prepared by the directors on the results and financial position of the company. The contents of the report

vary between jurisdictions and whether the company's share is quoted on a stock exchange. Usually the report includes statements of:
- The financial result for the period, often called a 'Profit and Loss Account' or 'Statement of Profit and Loss and Other Comprehensive Income'.
- The current financial position, frequently called a 'Balance Sheet' or 'Statement of Financial Position'.
- The company's cash flows, both inflows and outflows, called a 'Statement of Cash Flows'.

In addition, there will be a commentary on the business and the financial statements together with an indication on the company's future prospects.

The report may also be called the 'statutory report'.

Any Other Business (AOB) An item on the agenda for a board meeting which permits subjects which are not listed on the agenda to be discussed. This is often the last item on the agenda or the one just before 'Date of the next meeting'.

Articles of Association The document which describes how a company will be run (e.g., how to organise the Board of Directors' voting on decisions, or how to issues share certificates). This is sometimes called a company's 'constitution'.

Audit Committee A committee of the Board of Directors responsible for overseeing and advising on the content and standards of the company's financial reporting.

If the company's shares are quoted on a stock exchange, the minimum duties of this committee may be specified in the stock exchange's rules. The Board of Directors may add to these responsibilities.

Balance on the board It is generally considered that in order to make the best decisions (but still not necessarily right every time) a board should have a wide range of skills and experience represented at all board meetings. This range is often referred to as 'balance on the board' and sometimes as 'board diversity'. In this context 'skills' are knowledge gained by training and study, while 'experience' is knowledge gained through the performance of skills.

Balance Sheet A financial statement that reports the detail of a company's assets, liabilities and shareholders' equity at a specific time. The report is supported by Notes to the Accounts which include additional detail for some items. The figures are a mixture of historic figures and current valuations; therefore, care is needed when reading and using the figures. A Balance Sheet is often described as a financial statement giving a snapshot of what a company owns and owes. This statement may also be called the 'Statement of Financial Position'.

Best practice corporate governance guidelines A collection of advice and guidance on processes, ethics and policies to use in the governance of a corporation. Commonly expressed in general terms allowing the Board of Directors to apply the guidance in the most appropriate way for the company. The guidelines may be issued by government agencies and/or professional institutes.

Bicameral   A structure where the direction and control of an organisation is split between two separate boards, a Supervisory Board and a Management Board commonly found in Europe but occasionally, elsewhere.

Board advisor   An advisor is an expert in a particular subject retained to advise the board on specific issues. The advisor is not a director and cannot vote on any decisions. If the board has a permanent 'Advisory Board' or 'Advisory Panel', each advisor will be a member of this group of experts who gives advice to the board upon request.

Board assessment   A process where a company's Board of Directors reviews and assesses the way the board currently operates and achieves its objectives, seeking ways to improve. The process also includes the contribution each director makes to the board's review and decision-making processes. In some jurisdictions the board of a public company is required to have the performance of the company's board assessed periodically by an independent third party.

Board committee   The Board of Directors has the power to establish committees to assist them to deal with their range of responsibilities as well as to cover specialist areas. If a board committee is formed the whole board remains responsible for its functioning. The committees should operate by making researched and considered recommendations to the board who, after discussion, approve, modify or reject each recommendation. Most board committees keep the board fully advised by issuing reports or copying the minutes of their meetings to the board.

Board meeting package   The term used to describe the package of information needed by the directors so they can discuss the agenda items at a board meeting and make any decisions required.

Board of Directors   The governing board of a company elected by shareholders at the annual meeting of the shareholders. The Board of Directors are responsible for the future direction of the company and making policy.

Bonds   A bond is a loan made by an investor to a borrower which is evidenced by the borrowing company issuing a certificate. The certificate lists the interest the borrower will pay and when it will be paid, when the loan will be repaid and whether the loan is secured by any of the borrower's assets together with any other important terms.

Budget   A detailed estimate for a period of expected revenue, expenses and resulting profit together with capital transactions which together give an estimated future financial result and financial position. This process and document can be known by other names including a 'short-term plan'.

Business ethics/Corporate ethics   A statement of the acceptable rules and standards governing business policies and practices in relation to potentially contentious issues for example bribery, discrimination or fiduciary duties. To be used as a guide when deciding what is morally right or wrong.

Business review A discussion, prepared by the Board of Directors, of the figures in the company's financial statements in the context of the company's business operations together with an assessment of the company's prospects. The review may also be called the 'Operating and Financial Review'.

Cabinet responsibility The business convention and recommended practice where all directors must publicly support the Board of Directors' decisions regardless of their personal view or the effect on an area they manage.

Capital adequacy ratios A series of ratios which seeks to measure whether the company has sufficient resources to pay its debts as they fall due. The ratios may also be called 'solvency ratios'.

Capital expenditure Money spent by a company for an asset which is expected to have a useful life with the company, and which is longer than one year, and is expected to generate income or reduce costs. Also, money spent to extend the useful life of an asset and/or increase its earnings.

Capital structure The blend of equity (issued shares and undistributed profits) and borrowings (short-term, long-term and including leases) a company uses to finance its operations.

Cash Flow Statement A statement which summarises the actual cash - flows of a company for a period usually analysed into the cash flow from:
- Operating the business.
- Investing activities.
- Financing activities.

May also be called a 'Statement of Cash Flows'.

Chairman of the board A member of the Board of Directors elected to act as the person responsible for managing the business of the board, chairing all the meetings of the board and the company's meetings with shareholders, and for the effectiveness of the board.

Chief Executive Officer (CEO) The leader of the company's management team. The CEO may or may not be a member of the board. If the CEO is a member of the board an alternative title is 'Managing Director'.

Chief Financial Officer (CFO) The senior executive responsible for the financial aspects of a company. The CFO is responsible for identifying and managing financial risks, planning the company's finances as well as recording and reporting the company's financial figures. This position may also be called the 'Financial Director'. This position may or may not be on the Board of Directors.

Chief Governance Officer (CGO) A senior executive, who may or may not be a member of the Board of Directors, responsible for the company's governance systems and processes. In some companies the CGO is also the Company Secretary.

Chinese wall Effective actual or virtual separations within the organisation designed to block the flow of information between departments or sections in order to avoid potential conflicts of interest.

Civil Aviation Department (CAD) or similar The government department responsible for the regulation of aviation safety in a jurisdiction by deciding on the use of airspace, the operation of all airports and the licensing of local airlines as well as monitoring their financial position and management. There are a variety of other names for this department (e.g., Civil Aviation Authority or Civil Aviation Safety Authority).

Code of Conduct A set of rules, standards and proper practice to govern the behaviour of all the staff of a company, issued with the objective of ensuring that customers, staff, suppliers and all who deal with the company are treated fairly.

Commercially sensitive information Commercial information which, if known and used by competitors, would damage a company's business or commercial dealings. Examples of potentially commercially sensitive information are: the ingredients and formulae of a company's product, the profit or loss margins on individual products or routes or discrete business activities.

Communication The two-way process of sharing information, ideas and feelings so that all parties have the same understanding. There are many ways of sharing information, but communication is not successful until both parties have the same understanding.

Companies Registry A government office responsible for keeping a record of all the essential details for all companies.

Company Secretary An officer of a company responsible for keeping a record of directors' and shareholders' meetings, also for the legal records of the company including a register of shareholders. Other duties may be allocated when decided by the Board of Directors.

Conflict of interest A situation where an individual will, or is likely to, benefit from the result of a decision they are involved in making. Avoiding a conflict of interest is important for any company staff. There should be an established procedure for directors and staff to record potential conflicts of interest and for dealing with any conflict which may arise.

Continuing Professional Development (CPD) A continuing programme of education in the profession of being a director, often tailor-made for individual directors, to assist each director to stay up to date with legal and best practice guidance on corporate governance and also to increase the director's knowledge of the company and its industry.

Contribution A way of presenting the result for operating an airline's service or route where only the revenue and expenses, which can be clearly identified with each service or route, are shown and the surplus, often referred to as a 'contribution' is shown as contributing to meeting the airline's administration expenses.

Corporate governance There are many definitions for 'corporate governance', none of which are universally accepted. The definition used in this book is: the system of rules, laws, policies and practices which govern the operations of a company.

Current liability Any agreed amounts payable to third parties within the following 12 months.

Data Unprocessed facts and statistics which have been collected and are used as a source to be processed as information.

De facto director Someone who, although not openly appointed as a director of the company, frequently and routinely gives instructions to the board of the company on the operation of the company's business upon which the board routinely acts. Sometimes the de facto director owns or controls a significant number of the company's shares. De facto directors may be liable in the event of the company's failure. An alternative name is 'shadow director'.

Depreciation An accounting method of allocating the cost of a tangible asset over its useful life and is used to account for its decline in value. It is frequently considered to represent the reduction in an asset's value due to its use. The calculation of the amount of each year's depreciation charge is usually done using a formula. Regularly, a technical assessment must be made in case an additional exceptional charge needs to be made to reflect any reduction in value due to technological obsolescence or other factors. Depreciation is a 'non-cash' charge because money does not move when the accounting entry is made.

Derivative A marketable contract whose value is related to an underlying item. This might be an asset like shares, an index like 'Jet Fuel Price Index' or a financial transaction like an interest rate or foreign currency exchange rate. Sometime referred to as 'financial instruments'.

Director A person, working with others, jointly to accomplish the long-term success of a company, by setting goals, monitoring actual performance and setting ethical standards.

Directors and Officers Liability insurance (D&O insurance) An insurance policy usually taken out by a company with the premium paid by the company which reimburses a director or executive or the company for any loss or legal costs associated with defending any allegation of wrongdoing while a director or executive is performing their legal duties of managing the company. The insurance cover will not include deliberate criminal offences.

Directors' induction programme A plan which briefs each newly appointed director on the governance and operation of the company, including a review of the details of all the regulations, from all sources, on what is required and the penalties for not complying.

Dividend A distribution by a company to its shareholders of all or part of the company's profit. Generally, not all of the profit will be distributed, and the undistributed portion will either be re-invested in the company's business or held in reserve in case there is a year without profits. Dividends may be paid at any time during the year. Some companies pay a dividend at the end of each half-year or even every quarter. The dividend paid in respect of the company's result for the full financial year is usually called a 'final dividend'. Depending on the law of the jurisdiction in which the company is registered, the final dividend must usually be approved by a meeting of the shareholders.

Environmental, social and governance (ESG) This heading covers the effect the operations of a company have on the environment and the society in which it operates, also the policies and practices used in running the company. In most jurisdictions, companies are required, or encouraged, to report on these matters in the company's statutory reports to shareholders.

Exchange control permission The approval required from a government authority before money can be remitted out of a country.

Executive director A member of the Board of Directors who, in addition, has executive/management responsibilities within the company.

Executive officer A senior manager of a company holding executive authority. They may be a director on the board of a company or have the title of 'director', but not be a member of the Board of Directors, or neither have the title 'director', nor be on the board.

External auditors A firm of independent accountants employed to review and confirm the accuracy of a company's statutory accounts and express an opinion to the shareholders on their accuracy. In most countries an external auditor must be a member of an approved accounting organisation.

Extra-ordinary General Meeting (EGM) A meeting of the shareholders and directors called at a time between the company's AGMs to discuss matters related to the business of the company. The meeting may be called by the directors or by a minimum number of shareholders. The rules for calling and running the meeting will be detailed in the company's Articles of Association.

Fiduciary duty The duty of a person entrusted with the management of another's assets to act with the utmost care when dealing with all financial matters.

Finance committee A committee of the Board of Directors responsible for overseeing and advising on the company's finances and financial risk management.

Financial exposures The amount of potential loss from any investment. 'Investment' includes shares, bonds, deposits, bank balances, amount owed to the company, stocks. The company will try to reduce the financial exposure to the lowest level possible. Government issued bonds are thought to be risk-free, but some government bonds are high risk. Even investing in government bonds has the exposure to losses through inflation.

Financial Reporting Standards National accounting standards to be used for recording and reporting business transactions so that accounts are understandable and comparable. The standards are usually issued by a national accounting standards board authorised by the government. The trend is for National Reporting Standards to mirror International Financial Reporting Standards.

Financial restructuring Making a significant change to the amount of equity and debt a company has.

Fully costed Using an accounting method which calculates the total cost of providing a service or operating an airline's route. May also be called 'absorption costing'.

Governance Committee A board committee responsible for assessing the effectiveness of all aspects of the airline's corporate governance policies and practices, recommending changes and new policies and practices when appropriate.

Greenhouse gas (GHG) Greenhouse gasses (i.e., a gas which is thought, or proven, to contribute to the Greenhouse Effect – trapped radiated heat in the atmosphere).

Hedging A group of techniques used to reduce the probability of loss due to financial risks on, for example, fuel prices or interest rates. The techniques include entering into derivative contracts and/or having off-setting transactions. This area of risk management is usually the responsibility of the company's Treasury Department.

Hot topics Matters which are currently of great interest to shareholders and staff and society in general.

Independent non-executive director (INED) A director of a company who does not have any managerial function within the company, nor have any financial connection with the company other than receiving directors' fees. Sometimes referred to by the initials 'INED'. INEDs have the same responsibilities, duties and liabilities as other directors, and sometimes have a small shareholding in the company.

Information Data which is processed and presented in a form which is useful to the user.

Insider dealing Making use, by any person with special or confidential information, of the information, which is not generally available to others, in order to make a profit or avoid a loss.

Insolvent The inability of a company to pay its debts as they become due. In most countries there are legal consequences for the company and its directors if a company continues to operate while insolvent.

Institute of Directors An organisation founded to encourage high levels of competence, knowledge and integrity in all directors as well as the study and research into the development of good corporate governance. Not all of the organisations with these objectives are named 'Institute', they will adopt a name appropriate to their jurisdiction.

Internal Audit Department An independent department, frequently reporting directly to the Chair of the board, responsible for providing independent assurance that a company's risk management, governance and internal control processes are operating effectively and that transactions are being recorded accurately and in a timely fashion. In some organisations the department will report to, and support the operation of, a Risk Management Committee.

Internal controls Systems and practices put in place within a company to safeguard assets and prevent fraud.

International Financial Reporting Standards (IFRS) Accounting standards to be used for recording and reporting business transactions so that accounts are understandable and comparable across international borders. The standards are issued by the International Financial Reporting Board, an international standard setting board.

Investor An individual who, or organisation which, makes a financial contribution to a company in the expectation of making a financial return.

Joint-venture company A separate entity established by two or more companies or individuals to operate a defined business or achieve a defined objective, with the risks, profits and losses shared in proportion to the shareholding of each party.

Jurisdiction A territory which has the authority to make laws and over which its courts' authority extends and in which its decisions and interpretations must be followed. Usually the jurisdiction is a sovereign country, but there are exceptions (e.g., Hong Kong Special Administrative Region).

Key Performance Indicators (KPI) The principal set of measures a company uses to assess its progress towards meeting its strategic goals. The KPIs will include both financial and operating measures. There are books on how to relate KPIs to a company's strategies. The measures are sometimes called 'Success Measures'.

Lease A contract that gives one party the right to use an asset owned by another party for a set period of time in exchange for a consideration, usually 'rent', or a 'lease fee'. While the lessee continues to pay the consideration and keep the other terms of the lease, they may continue to use the assets without interference by the owner (lessor).

Limited company A legal entity, separate from its shareholders, which is a legal 'person' able to do all the things a person can do (e.g., enter into contracts or bring cases to court). It is responsible for its actions, finances, liabilities and commitments. The liability of its shareholders is limited to the amount they have paid for their shareholding (or have guaranteed in the case of a company limited by guarantee).

Liquidation The process, following the shareholders' decision to stop trading, during which all the company's assets are sold and all the amounts due to third parties are paid, after which any remaining surplus is distributed to the shareholders. If the proceeds from the sale of the assets are not sufficient to pay all the company's debts, each creditor will receive a proportion of the amount they are due, and the shareholders will not receive any money.

Liquid resources Any assets of the company which can be converted into cash quickly with little or no loss (e.g., deposits, government bonds).

Load factor The commercial definition is the ratio between the actual number of passengers carried compared to the maximum passenger capacity. This calculation becomes more complicated when an aircraft carries cargo as well as passengers (resulting in separate passenger load factors and cargo load factors). In addition, there are purely technical definitions.

Low cost carrier (LCC)   An airline whose approach is to minimise operating costs and relate as many costs as possible to the number of passengers carried. To achieve cost savings the airline may not provide some of the services offered by a full-service airline and may operate to and from less convenient airports. In addition, the airline may make a charge for some services (e.g., pre-boarding seat allocation, checked or carry-on baggage). Generally, the airline will only issue tickets for its own flights. This business model may also be called 'No frills' or 'Budget airline'.

Management Board or Committee   In a bicameral system, the Management Board is responsible for running the company in accordance with a plan agreed by the Supervisory Board and reporting to the Supervisory Board on the results.

In a unitary system the Management Committee's responsibility will be defined by the Board of Directors, but will be broadly the same as in the bicameral system.

The difference between the Management Board and the Management Committee is that the committee is not defined in law.

Management letter   A letter issued by a company's external auditors addressed to the company's Board of Directors, although the detail may be dealt with by the company's Audit Committee, which identifies issues and concerns found during the course of its audit of the company's accounting records and internal controls, which could lead to material errors or omissions in the accounts, or to fraud. The letter may also include suggestions for remedies or improvements. The contents must be taken seriously, and the board should respond to the letter indicating what action will be taken or explaining why no action will be taken.

Managing Director   The director who leads the company's management team. An alternative title is 'Chief Executive Officer', however the alternative title may also be used when the leader of the company's management team is not a director, so care is needed.

Measures and milestones   These are indicators of the progress and achievements of a project or strategy. Target measures and milestones will be set at the start of a project or for business strategies to provide a guide as to whether the project is in line with meeting its objectives. The measures relate to the achievements to be made and milestones when each achievement should be completed.

Memorandum and Articles of Association   A document combining a company's Memorandum of Association and Articles of Association. It describes the company, for example, by stating the company's name, and how it must be run (e.g., how share certificates are to be issued). Sometimes called a company's 'constitution'.

Memorandum of Association   The document which states the basic information about a company (e.g., the company's name, founding shareholders).

Mileage reward programme   Most airlines have mileage reward programmes which are designed to encourage customers to travel often with an airline (and similar arrangements for companies in other industries). This is done by awarding customers with miles or points or some other unit which can be used to pay for later travel. The basis upon which miles are earnt varies between airlines and can include the fare paid and/or the class travelled. The miles awarded may have an expiry date. There are standard ways to account for mileage reward programmes. Each airline tends to have its own name for its mileage programme (e.g., loyalty programme, frequent flyer programme).

Minority shareholders   A shareholder who does not own sufficient shares alone to pass a resolution at the company's AGM or EGM.

Nomination Committee (may also be called a Nominating Committee)   A board committee responsible for identifying potential candidates for appointment to the Board of Directors and to the senior management of the company. In addition, it will establish the range of skills and experience needed on the board.

Non-disclosure agreement   An agreement not to disclose to any third party any of the confidential information a director of the company has received during their appointment as a director. The agreement may have a provision that permits information to be shared with permission from the company.

Non-executive director   A member of the Board of Directors who does not have any executive/management responsibilities within the company.

Open tendering   A process which is open to all qualified and interested bidders, and which is generally advertised, and has an objective evaluation process which ensures a decision is made based on price and quality.

Operating and Financial Review (OFR)   A discussion prepared by the Board of Directors, of the figures in the company's financial statements in the context of the company's business operations together with an assessment of the company's prospects.

Organisation chart   A pictorial representation of an organisation's structure, showing the hierarchy, departments, sections, positions and titles.

Paradigm shift   A significant and important change of thinking in a business or industry which changes the current approach and practices to the way the business operates.

Policies   A recommended and approved course of action to be taken in defined circumstances in the operation of an organisation's business.

Practices   The actual methods of applying approved policies in the operation of an organisation. These are the usual ways of doing something.

Profit and Loss Account   An account or statement which has calculated the profit or loss produced by a company for a period. All of the company's revenue earned by operations is included together with all of the expenses related to generating that revenue. The statement may also be called the 'Statement of Profit and Loss and other Comprehensive Income'.

Provision for Bad Debts   An account in a company's Balance Sheet reducing the amount of trade debtors to allow for the chance that some debts may not be paid. This account may also be called 'Allowance for doubtful debts'.

Prudence To act with care and thought in the way a prudent person manages their own affairs, avoiding speculation, seeking long-term returns and with the need to preserve capital.

Publicly quoted company/Public company A company whose share may be bought and sold on a stock exchange at the prices reported by the stock exchange. The company may also be referred to as a 'public company'.

Quorum The minimum number of directors who need to be present in order for the Board of Directors to be able to conduct a board meeting and to make decisions.

Reconciliation The process which establishes the differences between two sets of records to ensure that any differences are understood and do not contain any errors. Frequently the differences are due to timing differences (e.g., differences between a bank statement and the company's record).

Register of Interests A register, usually kept by the Company Secretary, listing potential conflict of interest information for each director, senior manager and executive authorised to commit the company to transactions. The information should include any interest which someone might reasonably think may influence their decisions or actions. Examples of the information to be listed are: a shareholding in the company; shareholdings in other companies; or involvement in other businesses. In some cases, the interests of any family members will also be recorded if they might lead to a conflict of interests.

Regulator A legally appointed person who, or organisation which, is responsible for supervising an industry or some aspect of business activity.

Remit A task or responsibility formally assigned to an individual or a committee.

Remuneration Committee A board committee responsible for overseeing the process of setting the remuneration packages for all the company's directors, senior staff and the remuneration of all staff. If any of the remuneration packages include meeting certain targets, the committee must ensure those targets are sensible and reasonable, and align with the company's strategies. The principle underlying the operation of this committee is that no director or senior executive should set, or be able to influence, their own remuneration package.

Remuneration package All of the elements included in an agreement between a company and an individual of what the individual will receive in exchange for their services. The elements may include a salary, any allowances, bonuses, cash and non-cash incentives, contributions to pension schemes as well as non-cash benefits (e.g., a car). In addition, the agreement will detail any holiday entitlement and any limit on the length of the individual's employment.

Risk Management Committee A board committee responsible for identifying all the business risks involved in the airline's operation and recommending ways to manage them. In addition, monitoring the systems already in place to identify and manage flight safety risks by the Safety Committee, financial risks by the Finance Committee and operating safety and risks by an Operations Committee.

Safety Committee   A committee responsible for overseeing and advising on all flight and operating safety issues. The committee's terms of reference and operation must comply with the requirements of the airline's regulatory authorities which will often specify how flight safety is to be monitored and managed.

Secret profit   Making a profit from a transaction or accepting an advantage which is not disclosed, fully or partly, to the company. A director's duty under the law, and in accordance with the principle of dealing fairly, prohibits a director from making any form of secret profit.

Senior independent director (SID)   A member of the board elected by the independent non-executive directors to act as their representative to express any concerns they feel are not being fully considered at board meetings. A SID may also offer advice to the Chair and deal with any matters that relate to the Chair (e.g., conducting the Chair's annual performance assessment).

Shadow director   Someone who, although not openly appointed as a director of the company, frequently and routinely gives instructions to the board of the company which they usually act upon. Sometimes the shadow director owns or controls a significant number of the company's shares. Shadow directors may be liable in the event of the company's failure. An alternative name is 'de facto director'.

Shareholder   The owner, either an individual or an organisation, of at least one share in the issued capital of a company. The shareholder bears the business risk of the company and shares in profits and losses. They also have certain rights (e.g., to appoint directors to operate the company). A shareholder may also be called a 'stockholder'.

Small and medium-sized enterprises (SME)   This definition includes small companies and medium-sized companies. The latter is a company that is in the middle (i.e., neither small nor large). When used in conversation any definition for size can be applied (e.g., size of revenue, number of employees). When used in regulations or laws, a definition will be given. There is no internationally recognised definition.

Solvency ratios   A series of ratios which seeks to measure whether the company has sufficient resources to pay its debts as they fall due. The ratios may also be called 'capital adequacy ratios'.

Solvent (also Solvency)   The ability of a company to pay its debts as they become due. Therefore, a company must be confident it can repay its debts now and in the future. The inability to pay debts is called insolvent and in most countries, there are legal consequences for the company and its directors if a company continues to operate while insolvent.

Staff Committee   A committee responsible for discussing and making recommendations on all matters relating to staff; not only remuneration, but staff training. Also, the systems for dealing with staff and the two-way communication between the board, management and staff.

Staff cost add-ons The value of staff benefits other than salary (e.g., pensions, daily allowances).

Staff turnover The number or percentage of staff leaving the company and replaced by newly recruited staff during a stated period.

Stakeholders A person, group or organisation which is interested in or affected by the operation of a business.

Statement of Cash Flows A statement which summarises the actual cash flows of a company for a period usually analysed into the cash flow from:
- Operating the business.
- Investing activities.
- Financing activities.

May also be called a Cash Flow Statement.

Statement of Financial Position A financial statement that reports the detail of a company's assets, liabilities and shareholders' equity at a specific time. The report is supported by 'Notes to the Accounts' which include additional detail for some items. The figures are a mixture of historic figures and current valuations. Therefore, care is needed when reading and using the figures. A Balance Sheet is often described as a financial statement giving a snapshot of what a company owns and owes. This statement may also be called the 'Balance Sheet'.

Statement of Profit and Loss and other Comprehensive Income An account or statement which has calculated the profit or loss produced by a company for a period. All of the company's revenue earned by operations is included together with all of the expenses related to generating that revenue. The statement may also be called the 'Profit and Loss Account'.

Statutory reports (Statutory accounts) All the reports a company is required to produce, send to shareholders and lodge with government departments, on its operations. The reports include the statutory accounts which are the financial reports a company is required to prepare and send to shareholders by the law of the country in which it is registered or operates and/or the requirements of a regulatory organisation (e.g., a stock exchange). The accounts usually include a Profit & Loss Account, Balance Sheet (or Statement of Financial Position) and a Cash Flow Report, and cover a 12-month period. The statutory accounts will be included with, and cross referenced to, any other non-financial reports required by law or regulation. The contents of the reports are listed in the law or regulation and may include non-financial information.

Stock exchange A regulated market where company shares and other forms of investments can be bought and sold at prices agreed by the seller and buyer.

Stock exchange listing rules The rules of a stock exchange which every company seeking to have its shares listed and traded on, must follow. These rules detail the information required to be given before the shares are added to the share-list and traded. Also, the information and deadlines are provided for the information which must be given to the stock exchange and published periodically. The rules are likely to schedule the minimum corporate governance policies and practices the company must follow.

Strategy   A high-level statement which describes and defines the long-term objectives of the company or airline which has been approved by the company's Board of Directors. The statement of overall strategy is supported by strategies for each function in the company (e.g., marketing). It is the airline's view of the future. The overall strategy is sometimes called 'vision' or 'target' or 'goal'.

Subscribers   The original individuals who first formed the company. Sometimes referred to as 'founders'.

Succession plan   A systematic approach to identifying and developing potential leaders within an organisation who will be able to take over from the current leaders when they retire, with the objective of ensuring the organisation continues to operate and develop smoothly. The process is sometimes called 'replacement planning'.

Supervisory Board   In a bicameral organisation, a Supervisory Board oversees the working of a Management Board and is responsible for developing and agreeing strategies, policies and practices, reviewing and approving the Management Board's plans, monitoring the performance of the company and maintaining relations with the shareholders.

Terms of reference   A description of a committee's objectives, responsibilities, structure (membership) and the scope of the subject to be dealt with.

Test checks   To check a representative sample of transactions in order to verify the likelihood that most transactions have been correctly accounted for. Statistical analysis may be used to determine the size of the sample, but random sampling may also be used. If the sample shows irregularities, a more detailed review will be made until an opinion can be formed. The checks may also be called 'sampling checks'.

Tick the box approach   A term used to describe a narrow, bureaucratic attitude taken towards the subject of corporate governance rather than discussing the principles of good governance and applying them to the company's circumstances and needs. Taking a 'tick the box approach' to all aspects of governance means doing something solely because it is on a recommended list of things to do, thus apparently satisfying some rule or regulation or best practice guide. But in reality, the action is just following the form of good governance but not the content and meaning, and so is unlikely to achieve the desired objective of an open and transparent management system.

Total debt   The total amount of debt a company has, whether long-term or short-term, but the figure usually does not include trading debts such as the amounts due to creditors.

True and fair view   This means that after examining the statutory accounts the external auditors think the financial statements are substantially correct (i.e., free from material misstatements) and faithfully represent the financial performance and position of the company.

Unitary board A system where there is one Board of Directors which is responsible for developing and agreeing strategies, policies and practices for the company as well as running the company's operations. It is not unusual for some of the responsibilities of running the company to be formally delegated to a Management Committee.

Whistle-blower Policy A confidential process by which staff and suppliers can comment to directors or senior management on potential breeches of the company's Code of Conduct or any other wrongdoing by any staff.

Working capital A general term describing the funds used in operating a company's day-to-day business. In the company's Balance Sheet 'working capital' is the difference between the company's Current assets (e.g., stocks, debtors and its liabilities, for example, creditors). In many loan agreements there is a requirement that the borrower should ensure it always has 'positive working capital' (i.e., that its current assets exceed its current liabilities).

Wrongful trading The term used to describe a company's operation when it continues trading when the directors should reasonably realise that the company is insolvent. The term 'trading while insolvent' may also be used.

Yardstick An agreed standard for measuring an output.

Index

accounting and reporting standards 79; depreciation 79–80; policies 79; practices 80
advisor to the board 21–2
Advisory Board *see* advisor to the board
Advisory Panel *see* advisor to the board
AGM *see* Annual General Meeting
annual budget (budget) 38, 74
Annual General Meeting (AGM) 10, 19
annual report 9, 134–5
Any Other Business 39, 40
Articles of Association 34–5
associate company *see* subsidiary and associate companies
at arm's length dealings 103
Audit Committee 26–8, 32, 70, 80, 92

Balance Sheet 9, 69–70
benchmarking 125
bicameral board 5–6
board committees 25–32; in small companies 32–3; terms of reference 26
board diversity *see* balance on the board
board meeting 47–9; agenda 38–9; board meeting package 47–8
Board of Directors: annual work plan 38–40; assessment 43–6; balance on the board 22–5; business ethics 64–6; cabinet responsibility 17, 49; casual vacancy 128; communications 57–64; confidential papers 129; decisions 16; diversity 23; information 67; leaving 127–8; number of directors 23; possible business failure 97–8; reason for having a Board of Directors 4; reasonable person (reasonable man) concept 130; reporting to employees 137–9; reporting to shareholders 134–5; reserved powers 130; responsibilities 34–6, 130–1; role and duties 6–7; rules and guidance 34–6; skills and experience 23, 92; solvency 67–70
business ethics 64–6; suppliers and associates 65–6
business model 122–4; effect of changes 123
business review 56, 116, 122
business risk 82–4

cabinet responsibility 17, 49
capital expenditure after delivery 96–7
cash forecast 69, 75
Chair (manager of board matters) 19; assessment 44–5; separation of duties 19–20
checks and balances *see* internal controls
Chief Governance Officer *see* Company Secretary
Code of Conduct 64–6
communication 57–64; between directors 59; Communications Department 60; shareholders 59–60, 131–4; staff, reasons and methods 60–63, 137–9; stakeholders 63–4; strategy 50
company capital, increasing, 93–6; pre-emption rights 94; prospectus 95–6; types of shares 94
Company Constitution *see* Memorandum of Association and Articles of Association
company law 34
Company Secretary 8, 20–21, 36; part-time 21; sub-contract 21
company secrets 58
comparison with others 124–6; Key Performance Indicators 124
confidential papers 129
conflict of interest 40–42, 47; Register of Interests 41

Index

Continuing Professional Development (CPD) *see* director
corporate ethics *see* business ethics
corporate governance, benefits for a small company 86–7; compliance 13–14; changing with growth 85–6; definition 3; difficulty defining 3; keeping up to date 112–13; making changes 117–18; possible problems for a small company 87–91; problems for airlines 14–15; small companies 11–13; stable operations 91–3; suppliers 109–11; understanding and applying the principles 11–13, 15

depreciation 79–80; review 115
derivatives 114–5
director: alternate 18, 43; assessment 43–5; commentary in the annual report 9; Continuing Professional Development (CPD) 38, 46–7; cooperation with operating staff 5; definition 16–17; dissenting opinion 48, 128; duties 6–8; education 46–7; fees 89–90; financial connection 18; induction programme 36–8; legal duties 8; providing a service to the company 41; questions 72; reasonable care skill and diligence 4; removing from the board 127; rules and guidelines 34–6; shadow or de facto 18–19, 21; titles and types 16–18
Directors and Officers insurance 89

equity, increasing 93; or bonds 93–6; types of shares 94
ethics *see* business ethics
executive director 17
executive officers 17
external audit 32, 113–16; advice to the board and its committees 116–17; function 113; key areas 114–16; management letter 32, 116

fiduciary duty 4
Finance Committee 26, 28, 32, 81, 84
finances and financial exposures 81–2; reporting 81–2
Financial Reporting Standards 9; International 9; local 9

Governance Committee 30–32, 33, 91

independent non-executive director 18, 56; senior (SID) 20

information for the Board of Directors 67–72; finances and exposures 81–2; frequent routine reports 73–4; infrequent routine reports 74–6; other reports 76–7; price sensitive information 63, 131; volumes 71–2
insolvency *see* solvency
internal audit 118–20
internal controls 119–21; exception reports 120–21; using computer systems 120–21

joint-venture company 18; agreements 106–8; allocation of duties 6; governance 102, 106–8

Key Performance Indicators (KPIs) 52, 54–6, 86; types 55

Management Board 5
Management Committee 6, 27
material amount (materiality), definition 113
Memorandum of Association 34

Nomination Committee (Nominating Committee) 23, 26, 29, 33, 92
non-executive director 17–18

Operating and Financial Review (OFR) *see* business review

Profit and Loss Account 9, 73
profit plan *see* annual budget
prudence in finance and accounting 77–9

quorum 42–3; present at a meeting 42–3

reasonable person (reasonable man) concept *see* Board of Directors
relationship with staff 137–9
Remuneration Committee 26, 28–9, 32–3
reporting: by all sizes of company 11–12; shareholders 8–10; staff 11; stakeholders 10
reserved powers *see* Board of Directors
retirement plans 139–41
risk 82–4
Risk Committee 29–30, 33, 83; using technology 136

Safety Committee 27, 83
Salomon v. Salomon & Co Ltd, 1897 3–4
secondary capital expenditure *see* capital expenditure after delivery

Index

separation: of shareholders and the company 4; of directors and operations 4–5
shareholders: different objectives 9–10; own the company 4
small companies: benefit of independent non-executive director 18; governance 11–12
solvency: definition 67; protection 68–70; trading while insolvent 68
solvent 7
staff at board meetings 22
Staff Committee 27
stakeholders 10–11
Statement of Cashflows 9
Statement of Financial Position *see* Balance Sheet
Statement of Profit and Loss and Other Comprehensive Income *see* Profit and Loss Account
statutory report *see* annual report
stock exchange listing rules 35
strategy: changes in circumstances 54; communication 50; definition 50–51; developing and formalising 51–3; different names 50–51; early signs of failure 97; fails 97–8; measures and milestones 53; resources and planning 53–4; review 56–7; supporting strategies 51–2
subsidiary and associated companies: advantages and disadvantages 100–01; associate (associated) 103; branch office 101; governance 104–9; investment 104, 108–9; possible reasons to form 99–100; potential difficulties in different jurisdictions and cultures 106; subsidiary 101–2; types of entities 101–3
subsidiary company *see* subsidiary and associated companies
Success Measures *see* Key Performance Indicators
Supervisory Board 5
suppliers: governance 109–11
systems of control *see* internal controls

technology: Risk Committee 136; using 135–7
tick (ticking) the box approach 13–14, 39–40

unitary board 6

Whistle-blower Policy 64, 121; whistle blower 121
willing buyer and willing seller dealings 103

For Product Safety Concerns and Information please contact our EU
representative GPSR@taylorandfrancis.com
Taylor & Francis Verlag GmbH, Kaufingerstraße 24, 80331 München, Germany

www.ingramcontent.com/pod-product-compliance
Lightning Source LLC
Chambersburg PA
CBHW052128300426
44116CB00010B/1815